HISTORY OF IMPERIAL CHINA

Timothy Brook, General Editor

THE TROUBLED EMPIRE

CHINA IN THE YUAN AND MING DYNASTIES

Timothy Brook

THE BELKNAP PRESS OF
HARVARD UNIVERSITY PRESS
Cambridge, Massachusetts
London, England

Printed in the United States of America

First Harvard University Press paperback edition, 2013

Library of Congress Cataloging-in-Publication Data
Brook, Timothy, 1951-
The troubled empire : China in the Yuan and
Ming dynasties / Timothy Brook.
p. cm.
Includes bibliographical references and index.
ISBN 978-0-674-04602-3 (cloth : alk. paper)
ISBN 978-0-674-07253-4 (pbk.)
1. China—History—Yuan dynasty, 1260–1368.
2. China—History—Ming dynasty, 1368–1644.
3. China—Environmental conditions. 4. China—Economic conditions—To 1644.
5. South China Sea—Commerce—History. 6. Despotism—China—History.
7. Complexity (Philosophy)—Social aspects—China—History.
8. Commerce—Social aspects—China—History.
9. Commerce—Environmental aspects—China—History. I. Title.
DS752.3.B76 2010
951'.026—dc22 2010007195

CONTENTS

MAPS

FIGURES

INTRODUCTION

TWO dynasties ruled China between the middle of the thirteenth century and the middle of the seventeenth. The first was the Yuan, founded in 1271 not by a Chinese but by a Mongol, the Great Khan Khubilai, grandson of the world conqueror Chinggis (Genghis) Khan. The second was the Ming, created by the brilliant and ruthless Zhu Yuanzhang in 1368 and overthrown in 1644 by the next conquerors from the northern grasslands, the Manchus. This is a history of their dynasties.

For most Chinese, 1368 is a key moment in the history of these four centuries, for this is the year in which Zhu's indigenous rebel regime drove out the hated Mongols and re-established China as the realm that Chinese have learned to call "the fatherland." Historians outside China equally attach significance to the year 1368, but as the year that marks the onset of the late empire and the beginning of China's long approach to the modern world. The year 1368 plays a role in this book too, but differently: not as a hinge on which Chinese history turns, but as a brace connecting two parts. The revolt that brought the Ming to power did halt a century of Mongol rule, but it also ensured that the Mongol legacy would pass down through the centuries that followed. Together, the Yuan and Ming forged Chinese autocracy, reorganized Chinese society as a congeries of extended families, and restructured Chinese values to facilitate the concentration of commercial wealth.

Picturing China's transformation through the thirteenth to seventeenth centuries as a coherent arc of history was not the vision with which I approached the writing of this book. I started out by assuming that the Mongol century from 1271 to 1368 was a self-contained unit of time, a break in continuity from which the Ming recovered to set China on its course to the present: foreign became indigenous, Mongol became Chinese, black became white, or so I thought. The idea that the Yuan and Ming dynasties might be component parts of a single period arrived from an entirely unexpected quarter. In the course of reading through the four main genres of primary sources for the history of these two dynasties—the official dynastic histories, the court diaries or *Veritable Records,* the gazetteers that county administrations produced to record local affairs, and the commonplace books of essayists—I began to notice repeated references to natural disasters: famines, floods, droughts, tornados, locusts, epidemics, even dragon attacks. As I collected these references and arranged them over time, I found the two dynasties forming a single era that coincided with what climate historians working on other parts of the world have called the Little Ice Age.

What had been a warmer, wetter world became a colder, drier one. As it did, in China as in Europe, much else changed along with the weather. States and societies strengthened and polarized. Economies linked and commercialized. People were forced to come up with novel ways to explain what was happening around them and to them, to legitimize the new arrangements that contained their lives, and to justify the new modes of conduct they adopted to make their ways in the world. The world became global, and with it, China.

No one in the Yuan and Ming understood these changes in this way, for they experienced them episodically, often disastrously, as they were unfolding. To find the pattern to these episodes, in Chapter 3 I identify nine sloughs (rhymes with "boughs"), periods from three to seven years of intensely bad weather and large-scale human catastrophe. These sloughs did not decide the course of Yuan-Ming history, but they shaped life and memory during these dynasties as strongly as any other factor.

Buffeted by weather anomalies, troubled by the insistent presence of foreign traders in their offshore waters, some people clung to past precedents for guidance. Others cast those precedents aside to conceive of new ways to organize the world and find a place for themselves in it. This is why the Yuan-Ming period was a time of much confusion and a place of much disagreement.

To capture the vibrancy and variability of this age, I have tried as much as possible to narrate this history through the stories, paintings, and voices of that time. This was not difficult to do, for one thing that sets this period apart from earlier periods of imperial China's history is the sheer volume of reportage, most intentional, some accidental, that people of that time left for us to find. Eyewitnesses rarely see the bigger picture, but they furnish the details through which the past can come alive to us after a hiatus of so many centuries. Their ideas may not be congruent with ours, but their pleasures and their panics we can recognize.

I shall begin with panics: the moments when dragons erupted into their world.

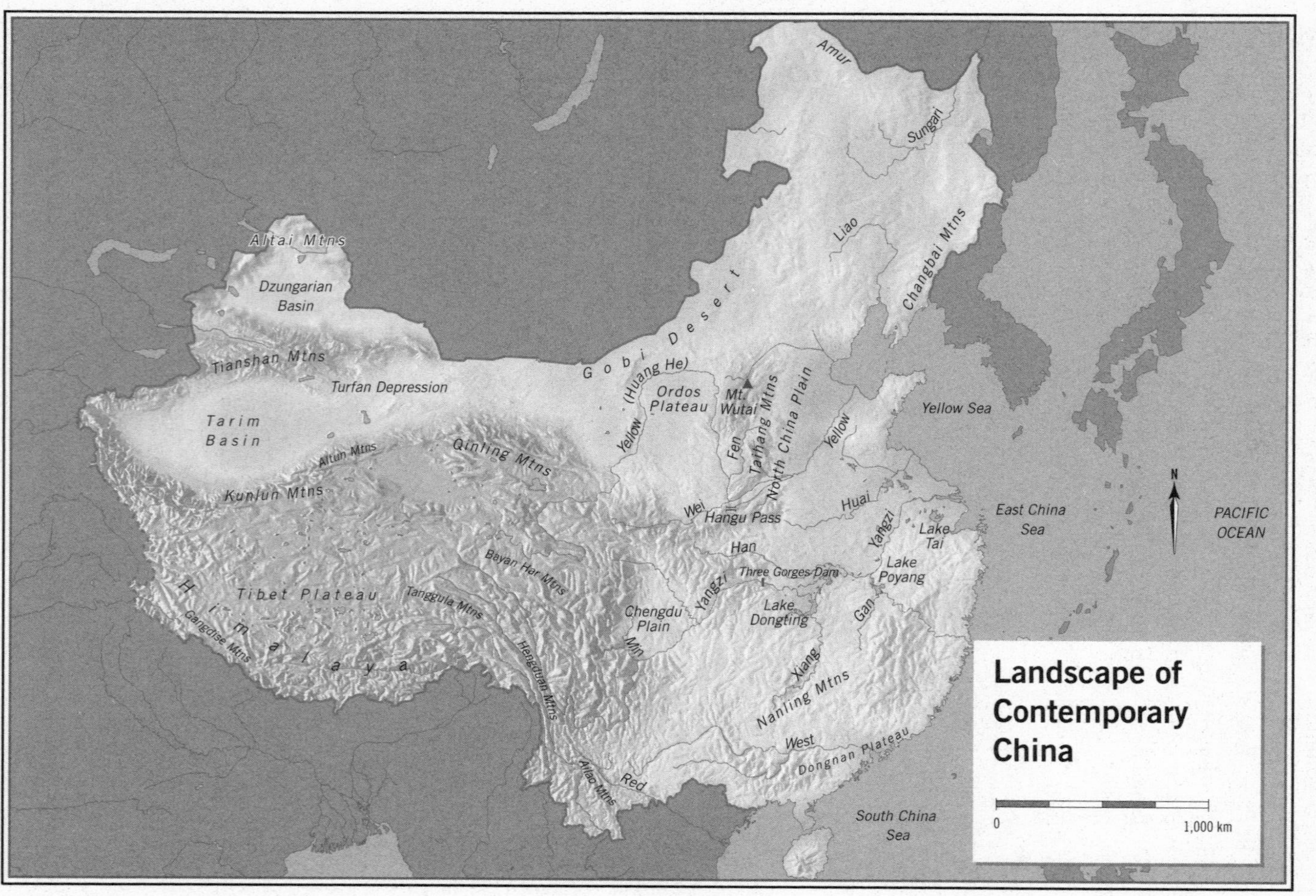
Landscape of Contemporary China
0
1,000 km
N
PACIFIC OCEAN
East China Sea
Yellow Sea
South China Sea
Amur
Sungari
Liao
Changbai Mtns
Gobi Desert
(Huang He)
Yellow
Ordos Plateau
Mt. Wutai
Taihang Mtns
North China Plain
Fen
Wei
Hangu Pass
Huai
Yangzi
Lake Tai
Han
Three Gorges Dam
Lake Poyang
Gan
Lake Dongting
Chengdu Plain
Min
Xiang
Nanling Mtns
West
Dongnan Plateau
Red
Ailao Mtns
Hengduan Mtns
Bayan Har Mtns
Tanggula Mtns
Tibet Plateau
Himalaya
Gangdise Mtns
Kunlun Mtns
Altun Mtns
Qinling Mtns
Tarim Basin
Turfan Depression
Tianshan Mtns
Dzungarian Basin
Altai Mtns

Map 1

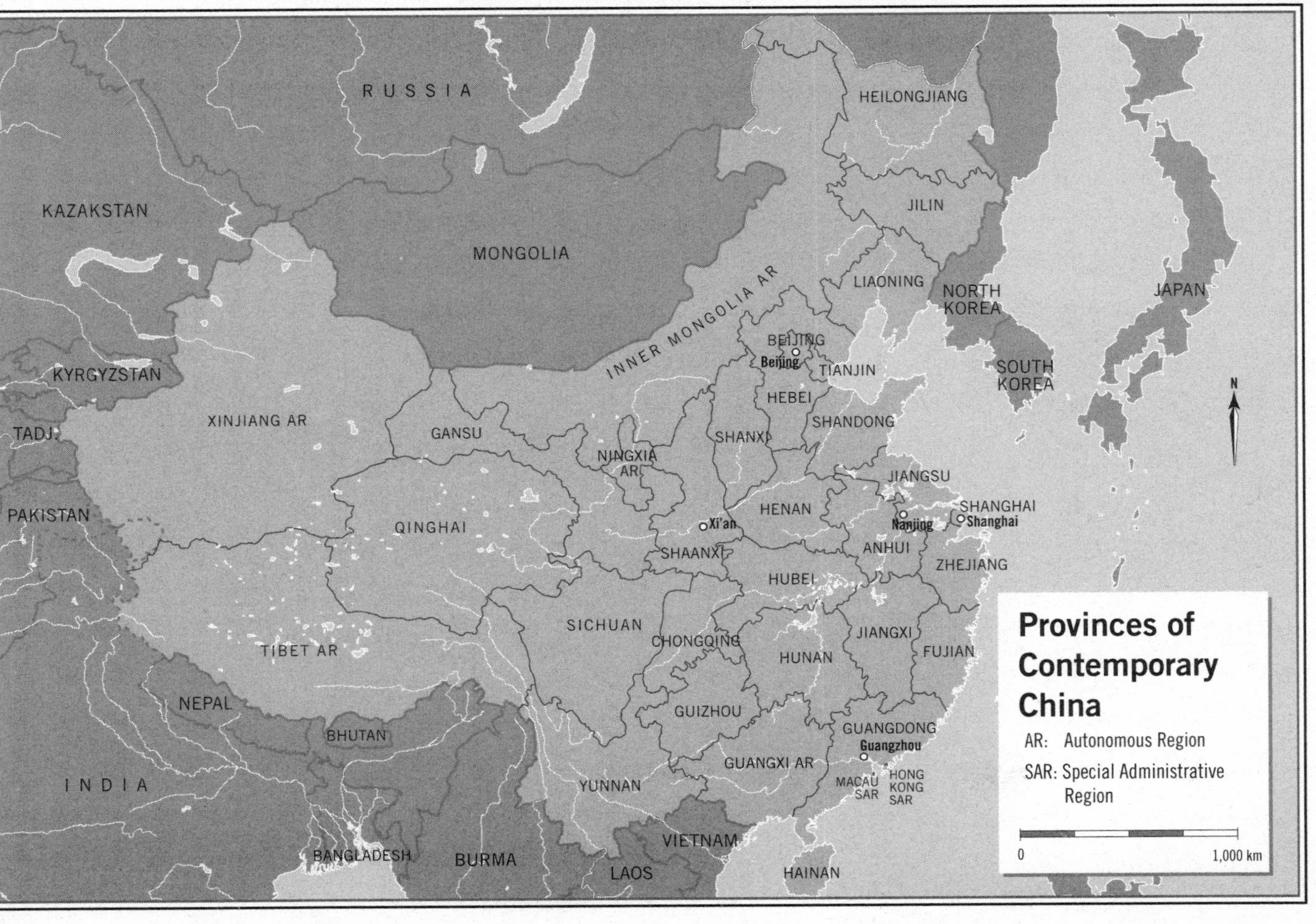
Provinces of Contemporary China
AR: Autonomous Region
SAR: Special Administrative Region
0
1,000 km
N
RUSSIA
KAZAKSTAN
MONGOLIA
KYRGYZSTAN
TADJ.
PAKISTAN
INDIA
NEPAL
BHUTAN
BANGLADESH
BURMA
LAOS
VIETNAM
NORTH KOREA
SOUTH KOREA
JAPAN
HEILONGJIANG
JILIN
LIAONING
INNER MONGOLIA AR
BEIJING
Beijing
TIANJIN
HEBEI
SHANDONG
SHANXI
XINJIANG AR
GANSU
NINGXIA AR
QINGHAI
JIANGSU
SHANGHAI
Shanghai
Nanjing
HENAN
Xi'an
SHAANXI
ANHUI
ZHEJIANG
HUBEI
SICHUAN
CHONGQING
TIBET AR
JIANGXI
FUJIAN
HUNAN
GUIZHOU
GUANGDONG
Guangzhou
GUANGXI AR
YUNNAN
MACAU SAR
HONG KONG SAR
HAINAN

MAP 2

I

DRAGON SPOTTING

THE FIRST dragon to show itself during the Yuan dynasty did so in 1292. This was the dynasty's twenty-second year, and two years before the death of its founder, Khubilai Khan (1215–1294). The dragon appeared at the edge of Lake Tai, the large body of water that occupies the heart of the Yangzi River delta and, like a heart, pumps water through the network of rivers and canals that crisscross this great alluvial deposit extending from the first Ming capital, Nanjing, down to the coastal port of Shanghai. As the dragon rose into the air, it unleashed a flood that submerged the fields clustered around the margins of the lake.[1] Rich farmland turned into marshy waste.

The dynasty's second dragon was sighted just a year later. It appeared on Chen Mountain, which despite its name is a modest hill 75 kilometers southeast of Lake Tai. Chen Mountain was marked with a shrine dedicated to the Dragon Lord dating back to the Song dynasty. It was called the Temporary Palace of the Dragon Lord, a term usually reserved for imperial palaces, since the lord, like the emperor, moved among many residences around the country. The shrine buildings having fallen into serious disrepair, the local magistrate decided to refurbish the place in the hope of currying favor with the Dragon Lord and getting him to bring rain to his parched county. Painters were at work at the site when, just before noon on August 25, 1293, thunder, lightning, and a sudden gust of wind announced the arrival of the dragon, and not just one dragon but two, the Dragon Lord and his young son. The pair of dragons revealed themselves to the awestruck painters below, then turned tail and disappeared into the clouds. As soon as they did, a foot of rain fell, ending the two-year drought that had parched the area.

Khubilai Khan died the following year. Three years after that, the dramatic visitation of the Dragon Lord and his son over Chen Mountain was quite overshadowed by a riot of dragons during a fierce rainstorm over Lake Poyang, the next major lake up the Yangzi from Lake Tai. Their aerial acrobatics whipped up surges that sent floods into the surrounding prefectures.

Dragons disappeared from sight for the next forty-two years. Their absence ended on July 29, 1339, when a fearsome dragon swooped down on an inland mountain valley in the coastal province of Fujian. The torrential downpour it released washed away over 800 homes and destroyed over 1,300 hectares of fields. Ten years later, five dragons burst once again from the clouds over the Yangzi delta, sucking spouts of ocean water into the air. Thereafter, dragons were spotted seven times in the seventeen years from 1351 through 1367. In that final year, the Yuan dynasty's last, there were two spottings. The first, on July 9, was in Beijing. A dragon emerged in a flash of light from a well in the palace of the former crown prince and flew off. Later that morning it was spotted in a nearby Buddhist monastery roosting in a locust tree, the bark of which was later found to be scarred and scorched. The second spotting occurred a month later, this time over Dragon Mountain in Shandong province, considered a potent site for praying for rain. During the August storm, the dragon appeared at the crest of the mountain.[2] Launching itself skyward, it loosened a boulder that rolled down from the summit and into local folklore. Eight months later, one of Khubilai's many great-great-grandsons was forced to abandon the Yuan throne and flee back to the Mongolian steppe. The foreign military occupation was over.

The Dragon Master

The Mongols' Chinese subjects had no difficulty interpreting these strange events. They looked at the growing litany of sightings through the final seventeen years of the Yuan dynasty, as rebels were rising on all sides, and knew that the dragons were heralds from Heaven announcing that the end of the Yuan dynasty was approaching. As an essayist who recorded the emanation of a white dragon during a cyclone on the Yangzi delta on August 10, 1355, observed in retrospect, "Every place the dragon passed became a desolate thorny landscape, withered and scorched." This was exactly how the landscape would look when civil war descended on the delta the following year. "Gazing upon this sight," he lamented, "the prosperity of former days felt like a dream."[3] A dozen years later, in 1368, Zhu Yuanzhang (1328–1398) emerged from the rebellions consuming central China. He "took flight like a dragon," in the standard phrase for becoming emperor, and founded the Ming dynasty.

Like Khubilai Khan, whom he revered as a great conqueror, Zhu desired to bend the world to his will. Between them, these two men did more to shape what China was during these four centuries, and what it would become thereafter, than any subsequent figure in Chinese history until Mao Zedong fought his way to power in the twentieth century. Khubilai's ambition had been to conquer East Asia. Zhu's desire was territorially more modest in scale. Instead, what mattered to him was transforming the battered realm he seized from the Mongols into a Daoist utopia, though it all too readily morphed into a Legalist gulag. Chinese today know him by his posthumous title Taizu, or Grand Progenitor, a standard honorific given to a dynastic founder. This was not the name by which he was known at the time, so I follow the conventions of referring to him either by his given name, Zhu Yuanzhang, or by his reign title, Hongwu, Surging Martial Power. Every emperor adopted a reign title to communicate his past accomplishments or future intentions; "Hongwu" reminded Zhu's subjects of his military achievements.

Before he became the Hongwu emperor, Zhu Yuanzhang was keenly aware of the dragons erupting into the Yuan realm. It was his metaphorical task to tame them, and Zhu was not the sort of person to leave metaphors alone. His first chance came early, in the fall of 1354, fourteen years before he founded his dynasty. He was campaigning up the Yangzi River west of Nanjing at the time. Drought had descended on the region. Local elders told him that a dragon could be seen from time to time in the nearby marshlands. They asked him to pray to it to stave off a full-scale disaster. "At the time I believed them, so I went to pray to it," Zhu wrote many years later. "After the third day, it in fact responded to my request" and rain fell. At the ceremony of thanksgiving, Zhu praised the dragon "for neither damaging nor flooding, adding merit to Heaven and Earth, succoring the people, and manifesting its efficacy before me"—just what he hoped his subjects might one day say of him. "On this occasion, the dragon has listened to Heaven's mandate and the spirits are all aware of this." Talk of Heaven's mandate—code for winning or losing a dynasty—

was a clear declaration of his intention to become emperor. Zhu ended the ceremony with this poem, which describes the dragon but sounds a lot like himself:

> Displaying its splendor, the cosmos is filled;
> Hiding away, nothing's revealed.
> The dragon's spirit controls the waters,
> The cosmic realm is pure and healed.[4]

Now that a dragon master was on the throne, the dragons did as they were expected to do: they withdrew from the human realm. Aside from a flock that appeared again in a storm over Lake Poyang the summer of his first year as emperor, no dragons disturbed Hongwu's reign. He was indeed the master.

Dragons of the Ming

The first dragon to erupt into the Ming realm did so in 1404, the second year of the reign of the Yongle emperor (r. 1403–1424). Several more were seen toward the end of this reign, the last unleashing an epidemic. As Yongle had come to the throne by usurping it from his nephew, who mysteriously died in a palace fire, people had good reason to suspect that Heaven was chastising the man. Not that anyone dared say that, for it would be high treason to suggest that the emperor should not be the emperor. Everyone kept quiet, and after his death the realm remained relatively dragon-free until the 1480s.

It was during the reign of the Hongzhi emperor (r. 1488–1505) that dragons began to be spotted with some regularity. Five of these sightings are noted in local sources, but just two found their way into the official *History of the Ming (Ming shi)*, compiled and published in the eighteenth century. The first of the official sightings, dated July 14, 1496, tells of a dragon that emerged from a soldier's scabbard during a lightning storm along the section of the Great Wall protecting Beijing. The second sighting, nine years later, reports an apparition swirling in a gyre of wind above the Forbidden City and then ascending into the sky at noon on June 8, 1505. "It looked like a man riding a dragon into the clouds," the historians reported.

These and the other Hongzhi-era sightings caught the attention of contemporaries. They may have been what inspired the skilled landscape painter (and notorious drinker) Wang Zhao to create what to my eyes is the most brilliant depiction of a dragon storm in the Ming, *Dragon Emerging* (Fig. 1). The sightings were a puzzle, especially to the Hongzhi emperor (Fig. 2). Widely admired as the first to get a grip on the problems of the empire in over half a century of incompetent and lackluster rulers, he cashiered incompetents, took personal interest in policy decisions, and managed court politics with skill.[5] How could Heaven be dissatisfied with such an emperor? Or was it the people, and not the emperor, that Heaven was putting on notice? The 1496 sighting at the Great Wall may have been what prompted him to send a eunuch attendant to the Grand Secretariat, his cabinet, to ask for information about dragons. The request sent the grand secretaries scurrying for someone in the administration who could provide such expertise.[6] (We shall meet the expert they found, Luo Ji, in the next chapter.) The last Hongzhi sighting on June 8, 1505—described as "a man riding a dragon into the clouds" over the Forbidden City—was not difficult to interpret, for it appeared at the moment of the emperor's death. Heaven was sending an emissary to collect a favorite son.

Being Heaven's creature, the dragon was the emperor's personal symbol. Only the palaces of the emperor and his direct descendants could be shielded from evil influences by a nine-dragon spirit wall. Only the imperial family could wear robes embroidered with dragons or eat off porcelains painted with their image—though the demand to imitate imperial fashion was so strong that embroiderers and kiln masters got around the prohibition by taking a claw off each of the dragon's feet, rendering them technically not dragons. In fact, the princes had to submit to the same demotion: in the only nine-dragon spirit wall to survive from the Ming, built in 1392 in the residence of the Prince of Dai in Datong, the dragons are all missing the fifth claw.

The association between rulers and dragons goes back to the mythic founders of Chinese civilization, who subdued the dragons living in the vast marshes of north China and converted the swamps into productive fields, turning the wild into the tame. Some emperors even kept dragons as pets.[7] The link was unambiguous, but its significance was two-sided. A dragon could display the authority of the emperor, but it could also signal that Heaven was unhappy with his rule. This is why dragons were record-worthy, even history-worthy. Signs from nature, they were fragments of a larger pattern that, if it could be read, might reveal the future course of national affairs.

The dynastic cycle furnished such a pattern. Heaven gave the mandate

Fig. 1 *Dragon Emerging* by Wang Zhao (fl. 1500). Wang has nicely captured the meteorological surprise that was thought to accompany a dragon's appearance. Palace Museum, Beijing.

Fig. 2 Portrait of the Hongzhi Emperor (r. 1488–1505). Notice the shoulder patches on his robe. The patch on his left bears the red image of the sun; on his right, the white image of the moon. Sun on the left and moon on the right: these symbols form the character *ming* ("bright"), the name of the dynasty. He is, of course, surrounded by dragons, the imperial symbol. National Palace Museum, Taiwan, Republic of China.

to rule to the man who proved he had it, by either seizing or keeping the throne. The logic is tautological, but was no less persuasive for being so. A founding emperor enjoyed Heaven's mandate and had no reason to expect dragon visitations, and anyone who claimed to see one was courting personal danger.[8] Dragons only came later, when the fortunes of a dynasty flagged and the prospect that the founder's family—his dynasty—might lose Heaven's mandate loomed. Hongzhi's dragonback ascent to Heaven—a story that court historians probably manufactured—showed that he enjoyed its favor, so in his case the dragons seemed to be a warning to the people to rally to their emperor, not a warning to the emperor himself.

When dragon spottings escalated under his successor, the Zhengde emperor (r. 1506–1521), the story changed. The first half-dozen years of the Zhengde era were dragon-free, until the night of August 6, 1512, when a fire-bright red dragon showed itself in the sky a hundred miles northeast of Shandong's Dragon Mountain. It circled ominously from northwest to southeast and then ascended into the clouds to a roll of thunder. It did no damage, however. Four years later, on July 7, 1517, nine black dragons appeared over the Huai River at the Grand Canal crossing, causing mayhem. As they sucked up water from the river, one of the boats was pulled up into a waterspout. The boatman's daughter was on board, but the dragon that sucked up the boat dropped her gently back to earth without harm. This bizarre scenario was repeated to worse effect a year later, when three fire-breathing dragons descended through the clouds over the Yangzi delta and sucked two dozen boats into the sky. The many people who died from the fall were outnumbered by those who died from fright. Over three hundred homes were destroyed, debris was scattered across the landscape, and red rain fell for the next five days.[9] These appearances were outdone eleven months later by a dragon battle over Lake Poyang. Dozens of dragons engaged on a scale that outdid the earlier displays in 1297 or 1368. Many an inundated island failed to resurface after the storm.

Everyone agreed that the Zhengde dragons were not signs of Heaven's favor. This emperor is remembered as the most irresponsible ruler of the entire dynasty.[10] Shen Defu (1578–1642) in his *Unofficial Gleanings from the Wanli Era (Wanli yehuo bian)* makes this interpretation crystal clear in an essay he entitled "Dragon Anomalies of the Zhengde Era." These dragons are not just general signs of a bad emperor but highly particular heralds of his bad judgments and bad end. Shen is able to align every sighting with a precise moment in that emperor's wobbly career, including the man's death. Zhengde died of a fever some weeks after falling drunk out of a boat while fishing in the Yangzi valley. Water being the dragon's signature element, Shen dared to suggest that dragons must have been responsible.[11]

From here on, dragons are regular accompanists to the Ming emperors as they stumbled from one constitutional or environmental crisis to the next. The reign of Zhengde's successor, the Jiajing emperor (r. 1522–1566), was heavily troubled by dragons, especially in the 1550s. Eighteen spottings can be precisely dated, but many more have been recorded that cannot. A Yangzi delta writer who collected local dragon stories from the Jiajing reign tells of one appearing in the home of a Hangzhou veterinarian, a second tearing up massive pine trees as it crossed Fang Mountain outside Hangzhou, a third and a fourth destroying dozens of houses outside Suzhou in a fiery heat, and a fifth roiling about over Hangzhou's scenic West Lake, where it toppled an iron pagoda, overturned pleasure boats, and smashed a Thousand Buddha Hall in a monastery to smithereens.[12]

The Wanli emperor's reign, from 1573 to 1620, was as dragon-prone as Jiajing's. Particularly spectacular was the second sighting, when 158 furious dragons tore out of the skies over the countryside west of Nanjing on August 29, 1586, leveling hills, destroying fields, and drowning thousands. Dragons continued to appear right through to the final reign of the Chongzhen emperor (r. 1628–1644). Two dragons were spotted in the autumn of 1643, the end of a brutal phase when it was impossible to pull

political, economic, military, and environmental crises apart. The dynasty fell the following spring.

Global Dragons

Chinese were not the only people to spot dragons in this period. So too did Europeans. The popular London science writer Edward Topsell devotes two chapters of *The Historie of Serpents* (1608) to dragons. Culling materials from numerous texts, including Konrad Gesner's widely reprinted *Historia Animalium,* Topsell jumbles everything he can find out about dragons into a barely coherent account. Dragons, he tells his reader, come in many different sorts, "distinguished partly by their Countries, partly by their quantity and magnitude, and partly by the different forms of their external parts." He allows historical examples to predominate, but eventually he brings the topic closer to home, declaring that "even in our own Country" many dragons have "been discovered and killed." His best-documented recent examples, however, are from the continent: France ("witnessed by many Learned and credible men which saw the same"); Lucerne on May 26, 1499 ("many people of all sorts beholding the same"); Germany in 1543 ("did bite and wound many men incurably"); and the Pyrenees ("a cruel kind of Serpent"). Scholars in Paris, he assures his reader, have dead specimens to examine. "It is said they were brought out of India."

Topsell knows some readers will doubt his stories, so he recycles what a German informant told Gesner when he provided him with information for his natural history. He insisted to Gesner "that he did not write feigned things, but such things as were true, and as he had learned from men of great honesty and credit, whose eyes did see and behold both the dragons, and the mishaps that followed by fire." Topsell offers the same defense, asserting that "this which I have written may be sufficient to satisfy any reasonable man, that there are winged Serpents and dragons in the world." He rounds out his case for their existence by weakly declaring that he would rather have his readers take him at his word than let them wait until a dragon actually appears in England, "lest some great calamity follow thereupon."[13]

Topsell's strong protests indicate that not everyone in Europe believed in the existence of dragons by 1608. Scholars in the preceding century were already disputing this matter. Topsell's declaration that his account of dragons did not "mingle fables and truth together" reflects the rising tide of doubt he was pushing against. On the other hand, his books sold well, suggesting that the belief was for many still intact.

The same split was emerging among Topsell's Ming contemporaries. To judge from the great encyclopedia, *All within Heaven (Tianzhong ji)*, which he compiled sometime after passing his metropolitan degree in 1550, Chen Yaowen reveals not a shadow of doubt that dragons existed in the distant past. This prolific scholar culled materials from a wide range of early texts to assemble for readers a complete understanding of all creatures, and in Chapter 56 he did the same for dragons. There we read that whereas water extinguishes human fires, it fuels dragon fires. We learn that dragons can see for 100 *li* (58 km). Like every creature, apparently, they are blind to something: "Humans cannot see wind, fish cannot see water, ghosts cannot see ground, goats cannot see rain, dogs cannot see snow"—and "dragons cannot see rock." We also learn that a dragon horn could be 6 meters long, this being the length of a red dragon horn presented to an emperor in 487.[14] Chen drew almost all his material from sources buried a millennium or more in the past. His Ming contemporaries were less certain about such knowledge, and some began recording their suspicions about dragons in the journals or "notebooks" *(biji)* that many kept. These journals are broadly similar to what English writers at the time called "commonplace books," which is the term I shall adopt for the Chinese genre. This is where intellectuals since the Song dynasty recorded the unofficial underside of things, and it is here that we find Ming scholars puzzling over dragons.

The Nature of Dragons

Most scholars of the Ming dynasty approached the task of determining the nature of dragons as a category problem. They did not ask what dragons were so much as to which categories they could be assimilated. Being airborne, fiery, and bright, dragons could be regarded as the most powerful manifestation of the yang half of the dualism of yin and yang (dark/light, female/male) that has long structured Chinese assumptions about the physical world. Yet the fact that they lurked in wells and other dark, watery places and brought rain and floods with a flick of their tails seemed to indicate an affinity with the yin half of the dualism. The core axiom of Chinese cosmology—that an extreme generates its opposite—did not quite resolve the problem, since dragons appeared to embody both extremes at the same time.

Serious scholars were stumped. Lu Rong's *Miscellany from Bean Garden (Shuyuan zaji),* one of the more informative commonplace books of the mid-Ming, will be a constant source of insights in this book. On the matter of dragons, however, Lu Rong (1436–1494) is uncharacteristically unsure of what he thinks. "They say that dragons can fly or lurk in the water, can be large or small." Having no basis on which to decide between these claims, he concludes that "their transformations are truly unfathomable: that much I can believe!"[15] Half a century later, Lang Ying (1487–ca. 1566), author of *Revised Drafts in Seven Categories (Qixiu leigao),* one of the more historically rigorous commonplace books, tries to sort out the yin/yang puzzle by doing what Ming intellectuals did well: sifting through every reference he could extract from earlier texts. This was also Edward Topsell's method, who was pleased to note that dragons "affordest and yieldeth so much plentiful matter in history for the ample discovery of the nature thereof."[16] Lang Ying would have agreed, and decided to focus on dragon births, guessing that reproduction must unlock the secret of their true nature. As it turns out, there was deep disagreement on this matter. Some people argued for embryonic birth, others for oviparous (shell birth). Hatching was more widely accepted—as Chen Yaowen confirms in his encyclopedia, *All within Heaven.*[17] Yet if that were the case, Lang asks, how is it possible that something born from an egg—birds at best, insects at worst—could have the magical capacities attributed to dragons? His curt observation—"the ignorant maintain that dragons are mysterious creatures capable of unfathomable changes"—suggests that, unlike Lu Rong, he will not accept the more fantastic claims made for dragons. In the end, however, Lang admits defeat. After going back and forth through the various debates on dragons (yang or yin? live birth or egg birth? warm-blooded or cold-blooded?), he lamely concludes that final answers will have to be deferred until "gentlemen capable of investigating the natural world" are able to come up with something more conclusive.[18]

The usual context for spotting a dragon was a fierce storm, so observers tried to work back from that evidence. Yet once questions started to be asked, storm reports came under suspicion. Lang Ying makes the point that storms impede clear observation. "When people of this age spot dragons, they are either suspended in the air or fighting, or crossing over or sucking up water," Lang points out. "This means that there is always wind and rain colliding, and thunder and lightning flickering in the gloom," making unambiguous identification difficult. Dragons also tended to keep their distance from humans. "When houses are destroyed and trees uprooted, the dragon may simply twinkle up in the clouds. All people see is the force of its coiling and twisting. They may want to see its entire body, but cannot."[19] As a result, the imagination takes over and fills in the blanks. To demonstrate his point, Lang dissects a suspicious sighting from the dragon-infested reign of the Zhengde emperor early in the 1510s, pointing out that witnesses jumped to the conclusion that they had seen a dragon after only the briefest glimpse.

Lang then turns to a sighting in Guangzhou (Canton). He was in the area at the time, though he does not specify whether he actually saw the creature himself. "One day after the morning tide had ebbed, a dragon fell from the air onto the sand," he writes. The immediate response was to defend the human realm from the animal. "Fishermen hammered it to death with the pikes they all carry. Officials and ordinary people gathered in large numbers to view the creature." Lang describes the creature as being "as high as a person and several dozen meters long. Its head, feet, and scales were just like they are in a painting"—life imitating art—"except that its underbelly was mostly a red color." Lang is satisfied with this sighting, closing the passage by declaring that "this is what can be called the proof of seeing." This dragon does not get him any further with the problem of how to analyze dragons, but at least it does confirm their existence. Logically, of course, the discovery of a creature that matches the traits that Chinese painters attribute to dragons does not prove that this was a dragon, and even less that dragons in general existed, but that was not the track along which Lang was thinking. The issue for him was never the existence of dragons; it was their properties. He had to eliminate the suspect sightings in order to keep mistaken information from creeping into his analysis of the zoological category to which dragons should belong.

A generation earlier, Lu Rong was not quite so confident of what could be ascertained from beached dragons. He recounts a similar story from the early 1450s regarding a marine creature that washed into Wenzhou Bay on the tide. Two dragons had battled in this estuary a century earlier (the second official sighting in the Yuan dynasty). Crowds descended on this marvelous animal not simply to see it with their own eyes but to harvest flesh from its carcass. Over a hundred amateur butchers were crushed when the animal rolled over on them and swam out to sea. Spotters at the time were uncertain whether this creature, which we would recognize as a whale, was a dragon, but they judged that it belonged to the dragon category. Lu is doubtful about the sighting, not the category.[20]

In the latter part of the sixteenth century, essayists appear to lose interest in probing the nature of dragons. They still report sightings, especially when a political interpretation seems to be in order, but they evince little enthusiasm for sorting out the issues that exercised fifteenth-century writers. The only extended inquiry into dragons I have found in late-Ming commonplace books is in Xie Zhaozhe's encyclopedic compilation of knowledge about the natural world entitled *Five Offerings* (*Wu zazu*). Xie devotes a fifth of the book to animals, and he gives dragons the opening thirteen entries. The first entry contrasts dragons, the most spiritually potent of creatures, with tigers, the most fierce: capture the one and you can rear it, but capture the other and all you can do is cage it. In the second he attacks physiognomists who claim that someone with dragonlike looks must have dragonlike powers. His rejection of physiognomy does not lead him to doubt other dragon lore, however. In his third entry, he explains that dragons are the most libidinous of all creatures. They will mate with non-dragons, producing hybrids that predictably share the characteristics of both parents. Half a dozen entries later he repeats the point, observing that "there is no creature with which they will not mate, which is why the types [of creatures they spawn] are uniquely numerous." Dragons will even mate with humans. Xie reports that rainmakers in the far south bring rain by exploiting this proclivity. They place a young woman out in the open as bait, and when a dragon descends on her, they prevent him from coupling with her. In his frustration the dragon ejaculates rain.

Xie nonetheless shares some of Lu Rong's skepticism. Sightings are a problem. Given that dragons always appear in the midst of rain and cloud, it is impossible for an eyewitness to see reliably the whole dragon, only parts of it. Xie is also skeptical about certain alleged facts about dragons. He notes the difference between human fire and dragon fire, then comments that "I really don't know whether this sort of thing can be believed or not." He is similarly doubtful about another popular claim that phoenixes like to eat the brains of dragons. "Phoenixes feed on nothing but bamboo seeds, so how could they possibly eat dragon brains?"[21] Doubt about any one aspect of dragon lore was not enough, however, to torpedo their reality. Dragons had occupied the capstone position in the order of creatures for as long as anyone could remember, and so they continued to do so during the Yuan and Ming. Even so, I suspect that, just as educated people in Europe were losing confidence in Topsell's account, so too late-Ming intellectuals were not entirely at ease with what people had thought they knew about dragons.

The surest evidence of dragons, which Xie mentions in his *Five Offerings,* were the skeletons that erosion exposed along river banks in the loess plateau west of Beijing. A landslide in 1636 at the village of Riverbend Bottom (Qudi) in southeast Shanxi province revealed a complete specimen. The teeth were over an inch wide, the skull measured five bushels in volume, and the clawed feet were four feet in length. Here was a dragon you could touch. The find was quickly broken up. The people of Riverbend Bottom were not curiosity hunters or amateur paleontologists. They had no interest in using fossils, as we do, to construct a history of the planet. Theirs was a far more practical concern, and one they shared with their European contemporaries: using dragon parts to cure illness. European medicine understood the medicinal properties of dragons to reside in their tissues (Topsell mentions fat, eyes, tongue, and gall), especially their blood.[22] According to Chinese medicine, however, the power of dragons was concentrated in their bones.[23] This is why the discovery of the bones excited local interest. A major drought had struck Shanxi three years earlier, and for the next decade the famines would go from bad to worse. With the famines came a sickness so severe that, in the words of a provincial historian, "the corpses of the starved stared at each other along the road."[24] By the time the dragon bones washed into view, the people of Riverbend Bottom needed all the medical help they could get.

Dragons as History

Dragons belong to Chinese history, but do they belong in this history? Yes, for the simple reason that serious historians of that era thought they did. If we turn to the Five Phases (or Five Elements) chapters in the two official dynastic histories, we find that the court historians have included dragons along with such abnormalities as plagues of locusts and snow falling out of season. When I first read these chapters, I found myself concentrating on the locusts and the snow and ignoring the dragons. Locusts contributed to famines, and snow falling out of season might be evidence of cooler temperatures. What were dragons evidence of?

Because historians at the time regarded dragons as record-worthy, we might gain something by trying to intuit what they meant for them, and therefore how they might mean something for us.[25] Whether the people of the Yuan and Ming believed in dragons is immaterial. They were observing phenomena that mattered to them, and if these events mattered to them, they should matter to us. The easiest course would be to put dragon sightings down to mass hysteria, but that does not get us very far. More interesting would be to take them metaphorically, as descriptors of extreme weather phenomena. A coastal dragon stirring up the ocean becomes a tsunami; a dragon tearing through a narrow valley marks a flash flood; a black dragon ripping up buildings and scattering rubble becomes a tornado; a dragon sucking up boatmen's daughters along with river water gets re-read as a waterspout; and so forth.

But reading dragons as weather, correct as that may be, runs the risk of missing the emotive or psychological—and political—impact of seeing dragons. The people of the Yuan and Ming grasped bad weather quite as well as we do, but when they saw a dragon, they saw more than bad weather: they saw a cosmic disturbance. Our inability to see dragons as dragons is our peculiarity, not a peculiarity of those who could. But are we in the twenty-first century as immune from overinterpretation as we tend to think? Don't we now think of bad weather as more than just bad weather, as a sign of global climate change—our own sort of cosmic disturbance?

Dragons were more than just animals, of course; they were terrifying creatures. The paleobiologist Stephen Jay Gould once remarked that dinosaurs excite our imaginations because they are "big, fierce, and extinct."[26] So too are dragons, except that for the people of the Yuan and Ming, dragons were very much alive. Indeed, China's last dragon was spotted over coastal waters in November 1905, just a few years before the fall of China's last empire, the Qing.[27] Spotting a dragon was an encounter with forces far greater than oneself. People did not just see dragons; they were riveted by them. Otherwise invisible creatures were showing themselves; the Heavenly realm was impinging on the human.

Are we so different? In recent years, Welsh farmers have reported seeing panthers around their farms. Panthers are not part of the Welsh ecology, and authorities deny the cats' existence. But many, inside Wales and out, are convinced the panthers are there. As the anthropologist Samantha Hurn has pointed out, the appeal of "symbolically powerful animals" reflects "the widespread human propensity to use non-human animals as definitional tools, or as metaphors for human actions."[28] Big cats provide those who claim to see them with a chance to expose what cannot be seen or even named. In the case of Wales, panthers "speak" for poor farmers who resent "English" regulations restricting fox hunting, which allow foxes to propagate unchecked and prey on poultry and domestic livestock. Farmers who feel powerless in the face of official regulations invoke panthers as a force of nature against the elusive power of the state.

This may help us understand why emperors had to assert control over dragons. Dragons reminded ordinary people of their vulnerability in the face of an unpredictable Heaven and an at times indifferent state. Those who spotted them could claim that these departures from normal were signs of the emperor's failure to attend to the needs of his people. Emperor Hongwu claimed the power to control them. Even the tragic last Ming emperor, Chongzhen, dreamt of seeing a black dragon coiled around a pillar in the palace when he was the crown prince—his own bid to claim that power and prove his qualification to be the next emperor.[29] Most emperors were not dragon masters and did not see dragons. In the Yuan and Ming, dragons showed themselves only to ordinary people. It was up to them to decide just what they meant.

Real panthers, if there are any in Wales, are indifferent to hunting regulations. They emerge from their hiding places to hunt for food, not to express political resentment. Even if there are none, panthers will continue to be spotted as warnings against the way things are, as flashes of insight into the way things ought to be. If there were real dragons in the Yuan and Ming, we would have to start all over to figure out how to incorporate them into a history that makes sense to us. But even if there were none, the storms they personified were real enough—and all the evidence people needed to know that dragons were lurking just beyond the edge of sight, ready to wash them away, but just as ready to chastise uncaring emperors for the tyranny and corruption that only increased their troubles. Had we been alive at the time, we would have seen them too. (Had we been English at the time, we would have known that Welsh dragons were the most dangerous type.)

Even if all we do is read dragon attacks as bad weather, that will still help us imagine a history of China that brings us closer to the past as people experienced it. As I shall argue in Chapter 3, the weather was indeed an active factor through the Yuan and Ming, shaping the trajectory of those four centuries quite as powerfully as the personalities and passions of the twenty-eight emperors who between 1271 and 1644 lurched from one crisis of legitimacy to the next. The nice thing about dragons is that they did not require one to make a distinction between bad omens and bad weather. They were both, each reinforcing the other.

The fearsome antics of dragons confirmed for those who lived through this period that these were difficult times, politically and meteorologically. They responded by crafting institutions and pursuing livelihood strategies to insulate them from these difficulties and keep their heads above water. As they did, much shifted from the way the world had been before. Autocracy and commercialization—to note two major themes of the period—were hardly unknown in the Song dynasty, but they were now present to a degree that was qualitatively, and not just quantitatively, different. Social practices diversified. Cultural production took new forms and served new purposes. Philosophers discounted many of the assumptions that had grounded Confucian thought. The cosmopolitanism of the Song dynasty was left behind. The Song was rhetorically present in the Ming as a good example of anything that seemed to require a good example (morals, institutions, habits), but no longer did it have the force of a model that anyone much tried to put into practice. The past was comfortable, but it was a fiction. The present required other ideals to make sense of the enlargement of private wealth, the cultivation of private emotions, and the alienation from state service that both of these developments encouraged. Especially during the last century of the Ming, the best and the brightest hotly debated which beliefs mattered and which did not. Was the prosperous and open age in which they had been born a better world, or was it a morass of busy profiteering and self-promotion that could only lead to moral and political ruin? Was the Way the way forward or the way back?

As the world within the Chinese empire changed, so too the world outside was changing. Merchants and sailors were weaving the Ming into trade networks around the South China Sea and, beyond that, to the Indian and Atlantic oceans. A global economy was taking shape, and the Ming, almost in spite of itself, was emerging as the key participant. A powerful convergence of environmental, political, and military disasters would conspire, however, to block that way forward. The end of the dynasty arrived in 1644 not from the ocean but, once again, from the steppe. But it was only the end of the Ming, not the end of the imperial system or the culture that supported it. That story, and the dragons who flitted in and out of it, would continue down to the twentieth century.

2

SCALE

IN THE fourteenth century, Europeans knew more about the Yuan dynasty than they had known about any other dynasty in China's long history. They knew it on the strength of the bestseller of the age, *The Description of the World*. This book told them about a realm more vast, populous, and prosperous than any part of Europe, ruled by "the mightiest man, whether in respect of subjects or of territory or of treasure, who is in the world today or who ever has been, from Adam our first parent down to the present moment."[1] The author, of course, was Marco Polo (1254–1324), and the ruler he adulated was Khubilai Khan.

The Polos, Venetian by the time Marco was born, hailed from the island of Korčula on the far side of the Adriatic Sea in present-day Croatia. Father Niccolò and uncle Maffeo broke free of the gravitational pull of the Mediterranean trading world and headed eastward in 1260, the year Khubilai was elected great khan of all the Mongols. The brothers reached his court at Karakorum on the Mongolian grasslands after five years of trekking and trading. They returned to Europe, then set off on a second odyssey in 1271, the year Khubilai founded the Yuan dynasty. On this journey they took Niccolò's seventeen-year-old son, Marco. That outing lasted twenty-four years. Its greatest legacy was Marco's book. For Europeans, one man's experiences of the Yuan dynasty became their introduction to Asia, and for centuries it defined what they thought China was.[2]

If Polo's book has a hero, it is Khubilai. "Everyone should know," Polo declares at the beginning of the core chapter, "that this Great Khan is the mightiest man." Everything linked with this ruler must be described in superlatives. His palace is "the largest that was ever seen"; the city around the palace is so densely populated that "no one could count their number"; the merchandise that can be bought there is more abundant than in "any other city in the world."[3] Hardly surprising that Polo earned the nickname of Il Milione, the Man of a Million Tales. This is the Yuan dynasty as Europeans believed it to be—a place as much of fantasy as of the real world, to which later writers such as Samuel Taylor Coleridge would return to fire their imaginations.[4]

Polo has been faulted for ignoring the feature we regard as symbolizing the vast size and might of the Chinese empire, the Great Wall. Frances Wood has even dared to ask whether Polo went to China at all. "Whether looking at a map of China today, flying over the north of China, or arriving on the Trans-Siberian railway, only someone who is seriously visually challenged could fail to notice the Great Wall and, indeed, be very impressed by it."[5] We look at this enormous feat of labor mobilization and see it as representing a polity on a geographical and political scale quite beyond European experience. Polo's failure to mention it when he enters Khubilai's realm in 1274 has led some readers to doubt his entire story. The objection seems to make sense, but if we put ourselves back into the thirteenth century, it is not so obvious that he missed something important. Polo says that he followed the Silk Route along the Gansu Corridor, entering the realm at Shazhou ("all idolators [Muslims], except that there are some Turks who are Nestorians"), traveling down to Ganzhou ("a large and splendid city . . . three fine large churches . . . many monasteries and abbeys"), then heading "southeastwards toward the countries of Cathay."[6] He does not notice the Great Wall at any point along the Gansu Corridor, for the simple reason that it was not there. Not until the second half of the Ming dynasty would anything worthy of being called a great wall begin to appear in the region.[7]

That there was no wall rescues Polo from the charge that he made it all up. The interesting point is that the Great Wall was not yet the symbol of Chinese might that it would later become. Straddling the sedentary world of China and the nomadic world of the Mongolian steppe, Khubilai would have been indifferent to walls. So too would the early Ming emperors, who imagined against the odds that the Ming would someday recover the steppe that Khubilai had once ruled. Later emperors gave up the idea, and gradually a wall was built in sections along the northern border, a defensive line separating nomads from farmers, Ming from Mongol, "Chinese" from "foreign." By the end of the Ming, hundreds of kilometers of wall had grown to thousands; but no, the wall didn't keep the nomads out (they came back in the guise of Manchus in 1644); and no, the wall is not so great that you can see it from space.

Unification

The Mongol mode of life was pastoral, and conquest was the logic of its rule. The tribe that stayed still, pasturing its animals on the same thin ecosystem, was the tribe that dwindled and fell under the dominion of others. The only way to survive was to move on, and so the leader who could lead his people to better terrain enjoyed a special charisma. Chinggis Khan pursued this logic right up to his death in 1227, pushing down onto the North China Plain, which was then under the control of the Jin dynasty of the Jurchens, a Tungusic people who had taken it a century before. Seven years later, the Mongols annihilated the Jin dynasty and began contemplating the conquest of the Song further south.

The Mongol conquest of the Song dynasty was delayed for five bitter years at Xiangyang and Fancheng, twin cities at the point of entry from the northwest into the Yangzi valley. Xiangyang finally fell in 1273, largely due to the technology of Muslim siege engineers.[8] The Mongols captured the Song court in Hangzhou two years later, though it would take them another four years to completely defeat remnants that moved ever farther south with members of the royal family in the hope of keeping the dynasty alive.

After taking Xiangyang, Khubilai turned his attention to Japan, which was supplying the embattled Song. He had already sent diplomatic envoys to that country in 1268, seeking to neutralize it as a Song ally, but that first mission was ignored, as were the second and third. Another means became necessary. It took the form of a combined Mongol-Korean force of 900 ships manned by 6,700 sailors and 23,000 soldiers that island-hopped its way across the Korea Strait in October 1274. It was a brutal attack, if the gruesome gesture of nailing the naked corpses of Japanese women to the sides of the ships is anything to go by. Japanese resistance was fierce and managed to stall the invasion long enough for a typhoon to strike, sinking a third of the fleet and drowning half the men. The invasion was called off.

Khubilai was able to complete his conquest of the Song without subduing Japan, but he dispatched a second, larger invasion in 1281. Leaky ships, inadequate supplies, hasty organization, and another cyclone spelled a second defeat. Out of it came the myth, manufactured in the nineteenth century, that Japan had been saved by a *kamikaze,* "divine wind." The term was revived in 1945 to honor the young men sent out as airborne suicide bombers against the final U.S. naval onslaught at the end of the Second World War.[9]

Having conquered the Song, Khubilai had to find a concept that justified Mongol control. He found it in the argument that the Mongols had earned the right to rule the subcontinent because they had reunified a realm that for centuries had been divided among the Song, Liao, and Jin dynasties. The idea of grouping all the territory he ruled within a single entity and calling it the Great Yuan dynasty was probably the doing of his most trusted Chinese advisor, Zicong. An ordained Buddhist monk, Zicong came into the Great Khan's service for a time in 1242 and returned to his service in 1249, becoming the principal architect of his regime.[10] Zicong understood that Khubilai could not win popular acceptance as the emperor of the Chinese people without submitting in some measure to Chinese traditions. One of these was to give the regime the status of a Chinese dynasty and place it within the long line of dynasties that had succeeded one another since the Qin unified the north in 221 BC. By declaring the founding of the Great Yuan, Khubilai claimed his place as the legitimate successor to the Liao, Jin, and Song dynasties. To set this succession in stone, he ordered scholars at court under his chancellor Toghtō to write the official dynastic histories of all three predecessors. It was a move to erase the longstanding Chinese distinction between *hua* (the civilized—and one of the names subsequently coined for "China") and *hu* (the nomadic peoples of the steppe). As the Chinese regarded the Mongols as *hu,* there was no point trying to persuade their subjects that they could claim *hua* status for themselves. Better to find a more inclusive concept, and that concept was unification *(yitong)*. Khubilai had gathered many peoples together under his rulership and made them one, and for that he deserved to claim that he was now Heaven's son.

The three dynastic histories of the Liao, Jin, and Song served to situate the Yuan in the line of dynastic succession by looking back. Khubilai aspired to do the same thing in the present, again on Chinese advice, by ordering the compilation of a national gazetteer: a comprehensive geography, administrative roster, and prosopography of all the places in the realm. This was a first. Gazetteers in previous dynasties were compiled only at the local level. The Yuan changed this practice. The *Unification Gazetteer of the Great Yuan (Da Yuan yitong zhi)* appeared first in 1291, followed by a second enlarged edition a dozen years later. This national publication set a standard for all subsequent dynasties. The Ming founder followed suit in 1370 by ordering the production of his own national gazetteer, though it would take decades for the project to be carried out. The order was repeated in 1418, and again in 1454 with greater urgency. The *Unification Gazetteer of the Great Ming (Da Ming yitong zhi)* finally appeared seven years after that.

The realm that Zhu Yuanzhang ruled was not the realm that Khubilai "unified." Zhu was forced to abandon the Yuan's Mongolian and Siberian territories, traditionally the zone of the *hu* nomads. Because the Yuan had claimed to unify the realm, however, the Ming could claim nothing less. Phrases such as "the unification of all under heaven" *(tianxia yitong),* "the unification of the present dynasty" *(guochao yitong),* or "the unification of ten thousand places" *(yitong wanfang)* clogged national discourse not just during Zhu's reign but for the rest of the dynasty.[11] It clearly mattered deeply to Zhu, to judge from a craven piece of doggerel he commissioned in 1370 entitled "The Great Unification Song":

> The Great Ming Son of Heaven mounted a dragon in flight,
> Opened the land and enfeoffed the princes to the left and right.
> From far off came the people to offer congratulations
> For bringing together all who dwell within the oceans:
> Warriors east, nomads west, foreigners north and south,
> Vietnamese and Mongols: all were brought onto the map.
> The imperial wind is blowing, we luxuriate near and far,
> Harvesting peace and bumper crops for a hundred millennia.[12]

The Ming was a large realm, but in every direction it was less extensive than the Yuan empire, even the Tang empire, for that matter.[13] The Yongle emperor (r. 1403–1424) aspired to return to Yuan borders by campaigning on the steppe and invading Vietnam, but on neither front could the Ming project its power for long. The entry on dynastic succession in the 1609 encyclopedia *Illustrated Congress of the Three Realms (Sancai tuhui)* tried to make the astonishing claim that "the Yuan rulers, who entered and ruled over China as an alien people," fell territorially short of earlier Chinese dynasties. "In the northwest they were unable to exceed earlier dynasties," alluding to the control of central Asia by another branch of the Mongol ruling house. "And the island aliens of the southeast did not all submit," referring to Japan's defeat of the two Mongol invasions. All this supposedly changed with the Ming. "When our imperial Ming received Heaven's mandate, it unified Chinese and aliens. The breadth of the territory that submitted extended all the way east of the Liao River, west as far as the desert, south beyond the maritime coast, and north into the steppes."[14]

This was just anti-Mongol rhetoric. By the mid-Ming, the regime had drawn well back from Yuan borders: according to the geographer Wang Shixing, five hundred kilometers on the north, two hundred and fifty on the northeast, a thousand on the northwest, and a thousand on the southwest. Of these zones, the most vulnerable to encroachment from the Ming was the southwest, which saw the slow, steady incursion of settled agriculture and state institutions throughout the dynasty and on through the Qing—a broad process of "absorption, displacement, and/or extermination" that the anthropologist James Scott has termed "internal colonialism."[15] The least tractable zone for expansion was the northern border, where the Ming eventually set up a buffer zone known as the Nine Frontiers and built the Great Wall to demarcate its outer extent.[16] "If one takes into account the fact that the Yuan rulers controlled the Gobi Desert," Wang observes, "then its territory at its fullest extent is not contained within today's realm."[17]

Networking an Empire

The ever-lurking consequence of grand scale is grand incohesiveness: too many locations scattered across distances too great for effective communication. This was a challenge that Chinese empires since the Qin dynasty had taken on, building networks of roads and canals across the length and breadth of the realm so that imperial messengers, officials, troops, and postal carriers could move speedily and at reasonable cost, and everyone else could follow in their wake.

Even before founding the Yuan and certainly thereafter, the Mongol empire developed astonishing communications networks. As it had to, for without the means to communicate across the vast distances, the Mongols would lack the means to control their far-flung territories. Marco Polo was greatly impressed by their system of land communication. "When one of the Great Khan's messengers sets out along any of these roads, he has only to go twenty-five miles and there he finds a posting station," he writes. "And you must understand that posts such as these, at distances of twenty-five or thirty miles, are to be found along all

challenge.

w/o communication lack control

the main highways leading to the provinces." Mounted couriers on urgent dispatch were expected to cover up to 250 miles in a day. It is, Polo assures his reader, "the greatest resource ever enjoyed by any man on earth, king, or emperor or what you will." Being from a small city-state in medieval Europe, he had never seen anything like it. "The whole organization is so stupendous and so costly that it baffles speech and writing."

Beneath the courier system and in parallel to it, the Ministry of War also operated a postal system to move routine government communications. This system relied on runners rather than mounted messengers. Polo describes these postal runners as wearing "large belts, set all round with bells, so that when they run they are audible at a great distance. They always run at full speed and never for more than three miles. And at the next station three miles away, where the noise they make gives due notice of their approach, another courier is waiting in readiness." In relay, these runners could traverse a distance in twenty-four hours that regular travelers would normally take ten days to cover.[18] Polo had good reason to be impressed. At the end of the sixteenth century, it took ten days for a letter to get from London to Paris, a distance of 400 kilometers. Back in the thirteenth, there would have been no guarantee that it would even arrive.

The Ming continued the Yuan system, though it cut back on some overland routes that entailed a heavy cost in horses.[19] According to a note in one late Ming commonplace book, the network of official roads extended 10,900 *li* (6,278 km or 3,900 miles) east to west and 11,750 *li* (6,768 km or 4,200 miles) north to south.[20] A modern scholar estimates that the total length of official land and water routes in the Ming amounted to 143,700 *li* (84,200 km or 52,300 miles).[21] The quality of the roads varied considerably from one place to the next, responsibility for upkeep falling on local magistrates and local budgets. The Ming founder minimized the cost of repaving the roads in Nanjing by declaring that steles (stone inscriptions) erected during the Yuan could be recycled as pavers.[22] As steles recording donations to Buddhist monasteries were most numerous, this meant marching into temples around the city and removing the tablets, a draconian move that many local magistrates would have found politically difficult to do. However, failure to maintain courier roads in one's jurisdiction was grounds for demotion, even dismissal, a threat that encouraged some local officials to keep them in good repair. For time was of the essence. Postal carriers were required to travel at a rate of 300 *li* (175 km or 110 miles) per twenty-four-hour period and were flogged twenty strokes for every three-quarters of an hour that a document was delivered late.[23] Couriers moved at a faster rate over longer distances, so their delays were punished in terms of the number of days they took above the time prescribed for a particular journey, at a rate of twenty strokes for every day lost.

Officials were allowed to use the courier system when they were traveling in the performance of their duties. The system was free for their use, though travelers were under fairly strict limits as to the speed at which they could travel and the services they could legitimately demand at the posthouses and courier hostels made available for their use. In the Yuan, for example, officials traveling by horseback were not permitted to cover a distance of more than three posthouses in one day, lest this pace strain the horses. Restrictions entered into the *Institutions of the Yuan Dynasty (Yuan dianzhang),* the official compendium of Yuan statutes, suggest some of the extremes to which official travelers thought they could resort. A substatute of 1287 required that officials arriving at a posthouse must deliver their horses directly to the courier stables to be fed before going drinking, not just dismount and leave the poor horses to collapse outside the tavern. Another substatute informed officials that they could not require posthouse officers to provide them with prostitutes—this as a result of a complaint filed in 1284 against a minor official who bribed a posthouse employee to procure him three prostitutes to sleep with, then repeated the demand the following night as well.[24]

The Ministry of War operated the courier system with military labor. For the convenience of official travelers, it produced a guide, *Network of Routes Connecting the Realm (Huanyu tongqu).* This cheaply printed handbook, first published in 1394, lists all courier routes in the country along with the 1,706 stations servicing them. Use of the system required a pass that specified the route and mode of transport. If the privilege included the use of courier horses, an image of a horse was stamped on the pass. If it did not and the official insisted on taking a horse anyway, he was liable to a punishment of eighty strokes. Like couriers, officials on assignment also had travel deadlines. Every route was rated according to the number of days the official could take to complete his journey (Map 3).[25]

For example, an official traveling by barge from Beijing to Nanjing had forty days to cover that distance; a day less if he was going only as far as Yangzhou, a day more if he was proceeding to Suzhou. Forty days was

MAP 3

also the time distance from Beijing to Yan'an in northern Shaanxi, as well as to Nanyang in southwest Henan. Cities in peripheral zones lay at a much greater distance. Chengdu, the provincial capital of Sichuan, was 145 days from Beijing, and Nanning in Guangxi province, 147 days. The city with the greatest distance rating, 149 days, was Chaozhou on the coast of Guangdong. After taking 113 days from Beijing to Guangzhou, official travelers had to snake their way east from the provincial capital over 1,155 *li* (675 km) of rough, slow terrain, which added another thirty-six days to the trip.[26] Had officials been allowed to travel by sea, the journey would have been much shorter, but they were not.

North/South

The separation between north and south was a national crisis during the Southern Song. Following the Jurchen invasion from the north, the dynasty found itself pushed out of its northern heartland and forced to split sovereignty with the Jin dynasty. The border ran along the Huai River, which flows west to east midway between the Yellow River to the north and the Yangzi River to the south. By reunifying the south with the north, the Yuan overcame this internal barrier. Physiographically, though, the distinction between north and south remained. Differences of climate, topography, food, architecture, and culture—even, it was believed, intelligence and personality—made these two zones different from each other. The north had the reputation of being dry, poor, and culturally backward; the south, just the opposite. Contemporaries were aware that the Huai valley was an ecological transition zone for agriculture. The land south of the Huai received enough rainfall to grow rice, which requires a minimum of 80 centimeters a year, whereas north of the Huai only wheat, sorghum, and other dry land grains could be grown. Wang Zhen, author of the Yuan dynasty *Agricultural Manual,* made this point in the fourteenth century, declaring the Huai River to be where grain agriculture divided between rice and millet. Another commentator writing two centuries later notes that parts of the Huai valley were suitable for growing rice, "so the price of grain is quite cheap there," while other parts were not, concluding that "this is where south and north meet."[29] Rainfall and warmth were the peculiar assets of the south, allowing for a more productive agriculture and greater investments in infrastructure, education, and cultural production that gave the south its dominant position.

Popularly, the dividing line between north and south was thought to be the Yangzi, not the Huai. "North of the Great River" (Jiangbei), that is, north of the Yangzi, was one world, and "South of the Great River" (Jiangnan), that is, south of the Yangzi, quite another. Ming essayists were particularly fond of expounding on this difference. Fujian writer Xie Zhaozhe in his commonplace book, *Five Offerings (Wu zazu),* divides them this way:

> Jiangnan has no locks, Jiangbei no bridges. Jiangnan has no outhouses, Jiangbei no cesspools. Southerners have homes without walls around them, but northerners cannot do that. Northerners build rooms without internal pillars, but southerners cannot do that. Northerners do not believe that southerners can build unbuttressed structures or raise them on stilts. Southerners do not believe that northerners have cellars that are ten thousand bushels in volume and lie hidden under the ground.[27]

The geographer Wang Shixing bases his more complex account on elaborate environmental differences:

> The southeast is rich in the profits from fish, salt, and rice. The central zone, Henan and Huguang are rich in gold and silver mines, precious gems, cowrie shells, amber, cinnabar, and mercury. The south is rich in rhinoceros and elephant horn, pepper, sapanwood, and the luxuries of foreign lands. The north is rich in cattle and sheep, horses and mules, wool and felt. Sichuan, Guizhou, Yunnan, and Guangxi in the southwest are rich in cedars and giant logs. South of the Yangzi there is abundant firewood, which means that people get their fire out of wood, whereas north of the Yangzi there is coal, so people there get their fire out of earth. In the northwest the mountains are high, so people travel by land and have no boats, whereas in the southeast the wide marshy lands mean that people travel by boat and rarely use horses and carts. Coastal southerners eat fish and shellfish, whereas northerners find the fishy smell repulsive. Northerners along the border eat yoghurt, the musty smell of which southerners loathe. People living north of the Yellow River eat peppers and onions, garlic and scallions, whereas people living south of the Yangzi fear pungent, spicy flavors.[28]

The core of the dominant south was the triangle of alluvial deposit in the lower reaches of the Yangzi River anchored at its northwest corner by the first Ming capital of Nanjing, at its eastern corner by the seaport of Shanghai, and at its southwest by the former Song capital of Hangzhou (Map 4). Also called Jiangnan, I shall refer to it as the Yangzi delta. This Jiangnan was within a single provincial jurisdiction under the Yuan, but Zhu Yuanzhang, the founder of the Ming, chose to split the delta between the two provinces of South Zhili and Zhejiang. Both his political instincts and his social conservatism induced Zhu to distrust the region. He preferred to divide and conquer. Zhu hailed from Fengyang prefecture in the Huai valley, the fault line between north and south, and was never comfortable among the elites of Jiangnan. Still, he was less of a northerner than the founders of earlier dynasties, and that caught the attention of observers then and thereafter. "The sagely emperors and enlightened kings of ancient times mostly came from the north, as did their ministers and advisors," notes the popular writer Wang Daokun (1525–1593), whereas "when the founding emperor of

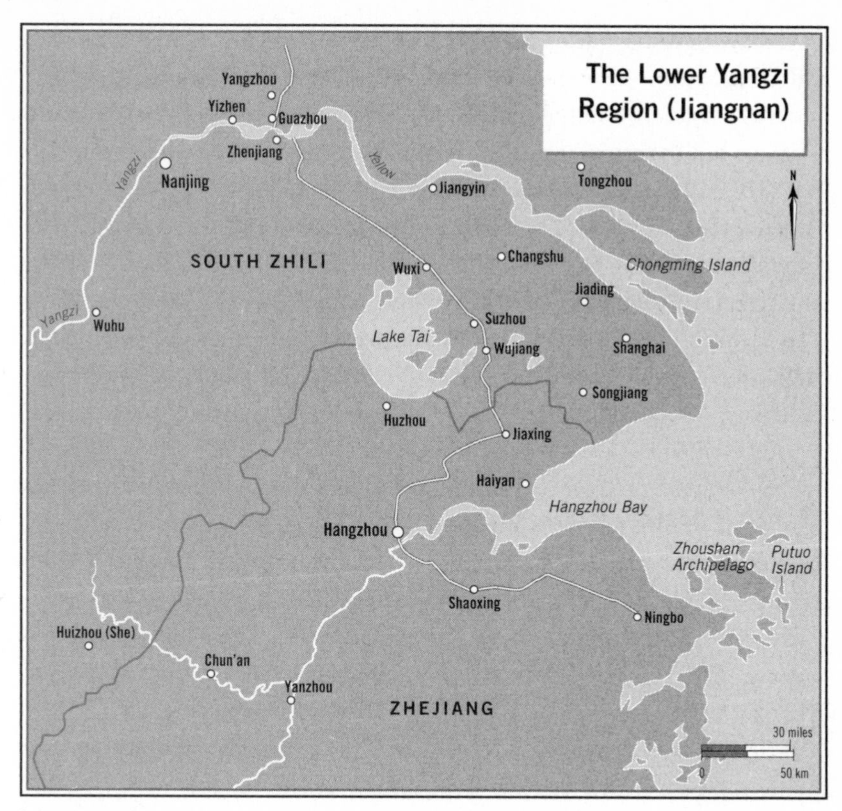

Map 4

our dynasty received Heaven's mandate and revived the realm, he arose in a southern principality: the south was the quarter in which his new realm dawned."[30]

Chinese culture might still look northward for its origins, but since the Song dynasty the rise of the south had been the motor of economic growth and the setter of cultural trends. In the longer duration, this was a recent change. As the geographer Wang Shixing liked to point out, "Jiangnan has enjoyed abundance and beauty for less than a thousand years." Only in the present, the sixteenth century, has "the entire region reached the peak of prosperity." And this logic of southward movement could well continue. "Who knows," he speculates, "when this might someday be true of Yunnan or Guangxi?"[31]

The Yuan dynasty constructed its administration across the divide between north and south, effectively perpetuating the distinction by relying on northerners whenever possible and distrusting the southerners who had resisted Mongol rule. Former Song and Jin elites brought together under the Yuan found themselves not reunited, as Yuan unity seemed to promise, but having to negotiate with each other across a divide of prejudice. This difference was experienced as a political and cultural tension between the northerners, who were dispatched south to administer the conquered territory of the former Southern Song, and the southerners, who were excluded from serving the new regime but obliged to negotiate with the new overlords. Southerners accused northerners of being uncultured and illiterate, while northerners regarded southerners as narrow-minded and self-righteous. Political accommodation under these conditions was difficult.

The Mongol practice of selecting officials through personal recommendation rather than examinations left a legacy of resentment among southerners, whom the Yuan regime severely excluded. After the Yuan fell, they wanted that imbalance righted in their favor, and saw the reinstallation of examinations as the means to do so. The examination system tested young men from all over the realm in a three-year cycle in order to grade them for government service. The cycle started in the counties, moved in the second year up to the provinces, and in the third was completed in the capital. Passing the county exams gave one the status of *shengyuan* (student officer); the provincial exams, *juren* (elevated person); and the national or metropolitan exams, *jinshi* (presented scholar). As the tests examined a common curriculum, the system contributed to the formation of a uniform national culture among the gentry. At the same time, however, it was generally thought that the system favored southerners, who traditionally had a better record of passing the exams than did northerners. Wang Shixing appeals to geography to explain this. North of the Yangzi, the landscape has a monotonous and uncramped regularity, so everyone shares a dull sameness and few strive for cultural distinction, whereas the more convoluted geological formations south of the Yangzi oblige people to live in more concentrated spaces and make them more competitive. Wang allows that not all great scholars and officials come exclusively from the south, but he feels compelled to conclude that since the reign of the Jiajing emperor (1522–1566), "the south has been overwhelmingly conspicuous in its success."[32]

The imbalance went well back before Jiajing. When Zhu Yuanzhang reinstated the examination system in 1370, he was conscious that southerners had been excluded from office and advancement under the Yuan, but he also liked the plain-spokenness of northerners and did not want to go too far in redressing the imbalance. The cultural advantages of southerners—more resources for education, more refinement of literary culture, a greater density of social networks supporting the attributes and attitudes sustaining scholarly production—meant that three quarters of the degrees conferred in the first national examination of 1371 went to southerners. Zhu was unhappy with this result and suspended the exams for a time. When he reinstated them in 1385, the ratio of southerners to northerners remained unchanged.

A problem became a crisis at the palace examination, a supplemental ranking exercise for the highest Presented Scholars, in 1397. All fifty-two graduates were southern. Zhu asked his chief examiner, Liu Sanwu, to reread the failed papers in the hope that worthy northerners had been overlooked. Alas for Zhu, Liu came up with exactly the same rankings. "In our selection there has been no distinction between southerners and northerners," he explained to the emperor. "It is just that south of the great Yangzi River there are many outstanding literati. Northern literati just don't compare to southerners." Zhu was so furious that he had two of the examiners executed (Liu was spared) and ordered a new palace examination. Not surprisingly, this time all sixty-one successful candidates were northern.

An administrative solution was finally imposed in 1425: 35 percent of the successful places in the national examination were reserved for northerners, and 55 percent for southerners, with the last 10 percent set aside for people from the ambiguous zone of the Huai River valley. These quotas were not applied to rankings, however, and as rankings determined who qualified for which posts, the impact of the reform on bureaucratic careers was muted. Of all the men who ranked first (winning the title of *zhuangyuan* or "optimus") in the palace examinations between 1370 and 1643, 80 percent came from the four southern provinces of South Zhili, Zhejiang, Jiangxi, and Fujian, in that order. Province of origin mattered statistically. If you hailed from these provinces, you stood a far better chance of scaling the ladder of success than if you came from the northern province of Shanxi, or for that matter from the southwestern provinces of Guangxi, Yunnan, or Guizhou, to name the three that did not produce a single optimus in the Ming.[33] Place of origin mattered culturally as well in terms of the resources available to students to prepare for and pass the examinations.

To improve northerners' chances in the exams, they were assigned all the places at the National Academy in Beijing. There was a National Academy in Nanjing, but it was for everyone else, which meant a much higher competition for places. Having a spot at the northern academy was additionally preferred because it gave a student better access to the Ministry of Personnel, which made all official appointments. Luo Ji (d. 1519), one of the scholars to whom the Hongzhi emperor turned when he needed more information about dragons, was a southerner who encountered the effect of affirmative action from the other side. From a county on the Jiangxi-Fujian border, Luo was an eccentric scholar whose unorthodox interests probably accounted for why he kept failing the provincial exams. Reaching the age of thirty-nine, he gave up that rat-race and bought himself the status of tributary student *(jiansheng),* which gained him a place in the National Academy in Beijing. The chancellor of the academy was the eminent statecraft theorist Qiu Jun (1420–1495), and he objected to a southerner being given a place. Luo persisted, and Qiu eventually exploded. "You may know a few characters, but that doesn't give you the right to be so obstinate!" To which Luo replied, "Perhaps so, but some people get appointed to the Hanlin Academy [the capital agency that advised the emperor on editorial and ritual matters] without ever having read a word!"

This should have been enough to get Luo sent home, but Qiu was intrigued by the man's bluntness. He permitted him to sit for the entrance test and discovered Luo to be a man of outstanding talent. Luo was allowed to take the provincial exam in Beijing in 1486, effectively jumping the long queue of southerners at the exams in Nanjing, and came in first.[34] It was a bit of an irony that the chancellor himself was from Hainan Island, as far south as you could possibly be born and still be a subject of the Ming.

Administrative Geography

The Mongols divided the territory formerly under the rule of the Jin and Song dynasties into nine administrative units, plus another three extending northward across the steppe.[35] The core of the Yuan realm encompassing the larger region around Beijing was called the Central Secretariat *(zhongshu sheng),* following the naming practice of earlier dynasties. This secretariat was also the central government's chief administrative agency. The rest of the country was subdivided into eight zones administered by eight Branch Secretariats: Henan-Jiangbei in the center,

Sichuan in the west, Jiang-Zhe in the southeast, Yunnan in the southwest, Shaanxi and Gansu in the northwest, and Jiangxi and Huguang in the south (Map 5).

The Ming started out using Yuan units, then revised them toward the end of its first decade. In 1376 the offices of the Branch Secretariats were abolished in favor of a triad of provincial agencies known as the Three Offices: the Provincial Administration Commission, the Provincial Surveillance Commission, and the Regional Military Commission. Their territorial jurisdictions were generally smaller than the Branch Secretariats had been. Three of the Secretariats (Shaanxi, Sichuan, and Yunnan) continued to use the Yuan boundaries. The Gansu Secretariat effectively disappeared, most of its territory escaping Chinese control with the Mongols. The Ming held onto Gansu's southeast corridor, attaching it to Shaanxi. The other secretariats were organized into smaller units (Map 6). The Central Secretariat was broken up into Shandong, Shanxi, and Beiping (later North Zhili or Northern Metropolitan Region, today's Hebei). The Jiangxi Secretariat was divided into Jiangxi and Guangdong, Jiang-Zhe into Zhejiang and Fujian, and Huguang into Guangxi and Guizhou. In addition, two new units were created by combining parts of different secretariats. South Zhili (or Southern Metropolitan Region, today's Jiangsu and Anhui) was formed by attaching the northern part of Jiang-Zhe to the eastern part of Henan-Jiangbei, and a new Huguang was designated by combining the western end of Henan-Jiangbei with the northern part of the old Huguang Secretariat. What was left of Henan-Jiangbei became Henan. These new units continued to be called *sheng* ("secretariats"), the word used today for "province."[36]

The fracturing of the Branch Secretariats into the Three Offices was a divide-and-rule tactic to forestall provincial officials from building independent power bases. In another way, though, the new system actually intensified administrative capacity at the provincial level by bringing three reporting systems together at a single node. The proliferation of provinces deprived the Ming of the capacity for interregional coordination that the larger secretariats of the Yuan system had made possible, but to address that problem, grand coordinators *(xunfu)* and supreme commanders *(zongdu)* were appointed, initially on an ad hoc basis in 1430, to deal with problems—primarily problems provoked by environmental degradation, such as flooding—that could be addressed only by crossing provincial boundaries. They were the regime's "environmental trouble shooters."[37]

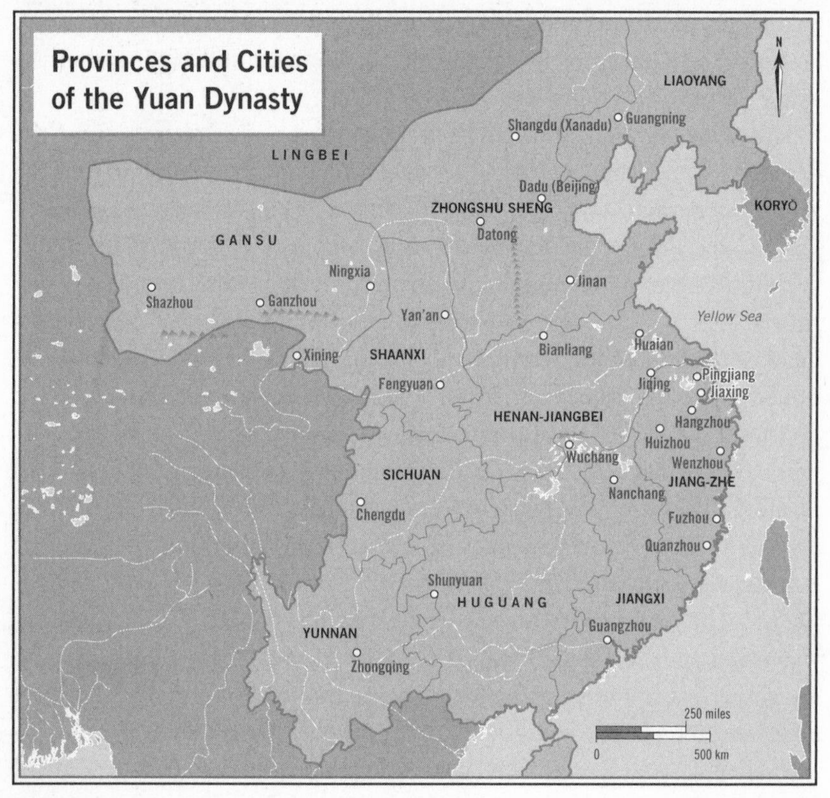

Map 5

Below the province were smaller administrative units (circuits and prefectures in the Yuan, prefectures and subprefectures in the Ming), which in turn were subdivided into counties, the basic unit of state administration. The Yuan had 1,127 counties at one point, the Ming 1,173, though these totals fluctuated as boundaries were revised. The county was the lowest unit to which the central government appointed an official. Each county had one magistrate, who was always a native of another province according to what was called the rule of avoidance, designed to prevent the retrenchment of local power at the expense of the center. The magistrate was responsible for overseeing the security and finances of anywhere from 50,000 to 500,000 people, depending on the size of his county. When the burdens on a magistrate became too great, a county could be subdivided and new counties formed.

New counties tended to appear in waves: fourteen in the 1470s, nine in

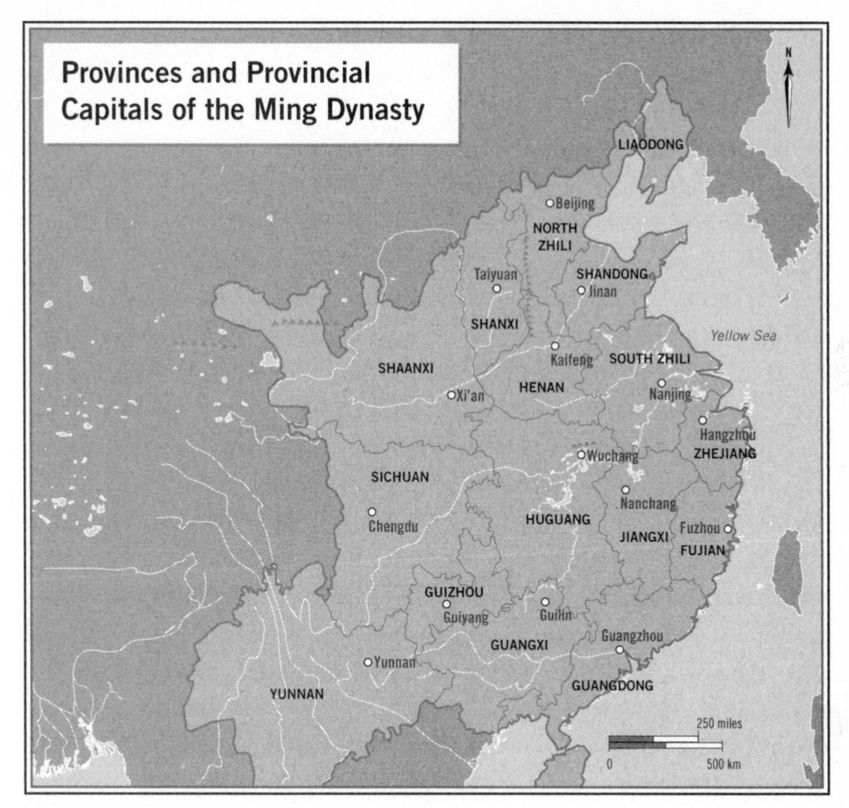

MAP 6

the 1510s, and eight in the 1560s. Many appeared in undersupervised peripheral regions, often in response to banditry.[38] In developed areas, counties emerged in response to economic growth. The town of Tongxiang south of Lake Tai, for example, was elevated to county status in 1430 in a major reorganization designed to improve fiscal operations in this densely populated region. Its surrounding market towns grew so dramatically on the textile trade that local leaders in the 1530s petitioned that each be granted separate county status. A resident of the commercial town of Qingzhen defended the proposal by explaining that the town had grown in size and prosperity because of its location on major transportation routes. "Its residents could not be under four or five thousand families," he observed. "Buddhist pagodas and Daoist monasteries," which depended on donations and were therefore a barometer of local wealth, "are jammed in everywhere you turn. The bridges and the market gates are regularly dismantled and rebuilt." County status would hardly do Qingzhen justice, he implied, for "the town really has the air of a prefectural seat."[39] The petition was turned down.

Haicheng county on the Fujian coast was created in response to both types of change. Haicheng was the site of Yuegang, Moon Harbor, the seaport for Zhangzhou that handled the bulk of the maritime trade linking the Chinese economy to Southeast Asia. The first bid for county status was in the early 1520s, but that attempt was scuttled when maritime trade was closed down in 1525. A second attempt in 1549 was turned down in a complex struggle at court over central control of coastal revenues. The third bid was first floated with an informal proposal in 1565, then formally proposed by petition the following year. The official who petitioned for status in 1566 phrased elevation as a solution to both a livelihood problem and a security problem. County status would mean more resources for defense against piracy, which would boost the town's import-export businesses. Disparaging the harbor's residents as difficult, indifferent to the law, and deep in cahoots with pirates, he also argued that county status would keep locals under better control. "There is stable revenue coming from the domestic boats and overseas boats, but that revenue is monopolized by local agents who are supposed to control the pirates," not by town administrators. "Notify the merchants that their payments should go directly to the officials" and there will be more than enough income for the new county to cover the administrative cost.[40]

The argument carried. Fujian got two new counties on January 17, 1567, and one of them was Haicheng. When the coast was reopened on a restricted basis that year, smuggling became trade, piracy became commerce, and Moon Harbor became Haicheng county.

Population

How many people lived in the Yuan and Ming realms? Both dynasties followed the tradition of conducting censuses, which were needed to determine how much labor was available to dragoon into state service when it was needed. Today, these documents provide us with information we need to understand the economy and society in the period. So we have many pages of population data—and yet the numbers seem so often wrong.

The population of the Song in the twelfth century exceeded 100 million, yet the first Yuan census in 1290 reported a population of only

58,834,711. The census takers understood that the real total was somewhat higher, acknowledging that "migrants living in the wilderness are not included in the total."[41] But an adjustment in the 1330 census raised the total only to 59,746,433. Did the transition from Song to Yuan entail a loss of 40 million lives? Did the Yuan occupation provoke a massive demographic collapse? Some historians have thought not, and have come up with proposals for increasing the figure by between 20 and 50 percent for 1290, producing a hypothetical population between 70 and 90 million. These numbers feel intuitively more reasonable for a realm the size of the Yuan, yet the troubles of the Yuan period must have depressed population in some places. We also know that many Chinese went unreported in areas where Mongol lords had enserfed them, causing them to disappear from the records.

The founding Ming emperor was eager to know how many people lived in his realm. On December 14, 1370, he notified the Ministry of Revenue that "despite the fact that the realm is now at peace, the one thing we don't know is the size of the population." He wanted a new census that recorded every member by gender and age (distinguishing preadults and seniors from full adults, since adults alone were liable for labor levy) as well as the amount of land the household possessed. This information was to be entered in duplicate onto a form, one copy of which was given to the household, the other inserted into population registers to be kept at the county *yamen*—the compound consisting of the magistrate's office, court, and residence.[42] These registers were known as Yellow Registers *(huangce)*. It was once thought that the name came from the booklets' yellow bindings. In fact, they were not bound in yellow. The name reflects the terminology of the four phases of life. A person became a child at three, an adolescent at fifteen, an adult at twenty, and an elder at fifty-nine. Unweaned children below the age of three were called "yellow mouths." Earlier censuses left young children off the official registers, recognizing that many would not survive the high natural rate of early childhood mortality and thus would never enter the tax system. Recording them was considered a pointless exercise. The Ming did not grant this exclusion—which is how the registers got their name. Thus "Yellow Register" properly means "a register that lists everyone including the yellow mouths." Not even infants could escape registration.[43]

The first census was taken in 1371, though some areas were left out. Contrary to the emperor's assertion that the realm was now at peace, not all of it was. A second census was ordered ten years later—as it would be every ten years, with only a few breaks, down to the end of the dynasty. According to the figures reported to the emperor in 1381, the Ming regime ruled over 59,873,305 persons living in 10,654,362 households. Ten years later, the total number of households grew by about 10,000, yet the population fell by three million. Something had gone wrong. The results were reviewed and a revised total was given in 1393 of 60,545,812. Adjusted or otherwise, these numbers are very close to the figures for 1290 and 1330.

The Ming mandated the updating of household data annually, and a new census every ten years. Magistrates dreaded the decennial count, known as a Big Compilation *(dazao)*, and usually just wrote in the last decade's figures, or switched a few numbers to give the appearance that a new count had been done. As a result, Ming census returns were as static as Yuan returns, giving the Ming an official population that fluctuated around the 60 million mark for the rest of the dynasty.

A truism of Ming administrators was that "an increase or a decrease in population attests to the strengths or weaknesses of conditions in the realm."[44] Increase meant prosperity, prosperity pointed to good government, and this is how a dynasty liked to be known. This conviction might have induced magistrates to report higher numbers, but that entailed an increase in the county tax quota, which no magistrate wanted, so the impulse was to keep the numbers as low as possible. Taxpayers in their turn tried to lighten their own burden by subdividing into households too small to be assessed for corvée. This trick pushed up the number of households and created the illusion of population growth while reducing the number required to meet corvée obligations. A satirical verse that was making the rounds in the Yellow River valley poked fun at all this:

> Barren soil along the river, the harvest not yet ready:
> New taxes are announced, and yet another levy.
> Every household subdivides, trying to evade them—
> And officials mistake the whole thing for a growth in population![45]

If locals could hoodwink the officials, what are we to do? Skeptics dismiss all later figures after 1393 as administrative fictions hiding real growth, but disagree on how large that growth was. They prefer to raise the starting figure by 10 percent on the assumption that the censuses could not have captured more than 90 percent of the population. Setting an annual rate of increase at three per thousand, they come up with a population in 1600 of 150 million—almost three times the size of the 56 million actually reported in the Big Compilation year of 1602. Ultra-skeptics want to push the numbers even higher by adopting a slightly higher starting population and applying a higher rate of annual increase (five to six per thousand), which drives up the population of the Ming in 1600 to as high as 230 million.[46] These proposals have provoked a reaction among those we might call the statistical fundamentalists. They doubt the wisdom of tossing out the figures we have for the figures we want. Sticking as closely as possible to the official censuses over the next two centuries, and working with a much lower annual rate of growth of 0.4 per thousand, they come up with a population in 1600 approaching 66 million.

The skeptics, ultraskeptics, and fundamentalists have thus given us three population estimates for 1600: 66 million, 150 million, and 230 million. These numbers have interesting consequences, for each estimate implies a different subsequent history. The Qing recorded 313 million in 1794 and an estimated 430 million by 1840. Which number we choose for 1600 affects how we interpret these later numbers. If we accept the fundamentalists' figure of 66 million, then eighteenth-century growth was spectacular: an annual growth rate close to 8 per thousand. If we adopt the ultraskeptics' 230 million, then the spectacular growth in the eighteenth century becomes a statistical illusion. The middle estimate, 150 million, smoothes out the long-term curve, suggesting a reasonably constant rate of growth of around 3 per thousand across both dynasties. I prefer the middle estimate.

Migration

Many people of the Yuan and Ming did not stay in the places they were born. Some were always on the move, whether by trade or necessity. Sometimes it was the state that moved them. When Zhu Yuanzhang expropriated wealthy farms on the Yangzi delta, he relocated some of their owners to the capital in Nanjing where he could keep an eye on them, some to his home prefecture in the Huai valley, and some to the depopulated North China Plain. The Yongle emperor's plan to move the capital to Beijing prompted further forced relations onto the North China Plain, some of them involving tens of thousands of households. Over the fifteenth century, the metropolitan province around Beijing, North Zhili, increased its share of national population from 3 to 7 percent.

Some who moved did so at the behest of the state, but most were economic migrants who traveled through private networks rather than public programs. One of these in the north was anchored in the shade of an old locust tree in the town of Hongdong in Shanxi. This spot was an assembly point for residents of this heavily populated province (which had remained relatively insulated from the Yuan-Ming transition) who were looking for land elsewhere. Hongdong, which straddled the Fen River corridor running up the great seismic fault that angles through Shanxi, was ideally situated for this service. Anyone heading out of the province could not avoid passing through Hongdong. But many purposely traveled there to join group migrations going out of the province. One enthusiastic local historian has tracked down evidence of out-migrating families ending up in roughly two out of every five counties throughout the realm. Some hailed from Hongdong itself, though many were from elsewhere in the larger Shanxi-Shaanxi region. Four fifths of the emigrants moved east onto the North China Plain, while the rest scattered to every other province in the realm.[47]

These emigrants kept the memory of their Shanxi/Hongdong origin alive. In their genealogy, the Wangs of Lotus Marsh record their descent from founding ancestor Wang Bosheng as a story of emigration:

> Our founding ancestor was originally registered in Wooden Raft Village, Old Crane Gulch, Hongdong county, Shanxi. He was one of four brothers, all of whom felt that, having been born and grown up there, they would end their days there. Little did they suspect that in April 1370 the emperor would order people to relocate to the east. The brothers desired that our founding ancestor should stay on the old property, pursuing his work and growing old there, but he and his second brother decided to obey the directive and move east. Crossing passes and mountains, rivers and seas, their trek was wearisome in the extreme. They traveled with the moon and stars lighting the way; the winds and frosts they faced were too bitter to describe. Our ancestor ended up in the southeastern corner of Dongming county in Daming prefecture [in North Zhili]. Three kilometers from the county seat he found a delightful spot and chose it as the place where he would settle, so he measured the land for a house and built a compound for himself and his family.[48]

The "delightful spot" was called Fendui, Dunghill Village, which meant that the family became known as the Dunghill Wangs. They later adopted a more polite choronym, Heze, Lotus Marsh, so that by the time they compiled the lineage genealogy in 1887, they were calling themselves the Lotus Marsh Wangs.

In national terms, the greatest outflow of people during the Ming was from Jiangnan. The three core provinces of South Zhili, Zhejiang, and Jiangxi accounted for half the national population in 1393; by the mid-Ming their share was approaching a third.[49] This huge redistribution was partly due to absolute growth in other provinces, but it was also fueled by westward migration, first to Jiangxi, thence to Huguang, and finally to Sichuan and Yunnan. Already by the 1420s, southern Huguang was inundated with economic refugees from further east, some of whom shaved their heads to appear to be monks on pilgrimage.[50] A central official who was sent to this region to coordinate the relief of a province-wide famine in 1509 was struck by the number of "out-of-province, out-of-prefecture, out-of-subprefecture, and out-of-county people" he found, almost all of them economic migrants from further east.[51] Jiangnan remained the densely populated core it had become in the Song, and it continued to shape social norms, economic practices, and cultural trends throughout the dynasty. Increasingly, though, its descendants could be found countrywide, sometimes eking out their lives on marginal fields carved out of hillsides, often dominating the local societies into which they moved, deploying the schemes and resources that came with their Jiangnan background.

The Administrative Matrix

Censuses and tax quotas, fixed residence and migration, aggregation and subdivision: these all show the state devising every possible mechanism to register and control every subject of the realm. What made this control possible was a system of interlocking administrative units extending downward from the county, the lowest unit to which officials were centrally appointed, into the countryside and touching every household. No life was possible that was not lived within state units—or that was the goal.

The Yuan employed a variety of local territorial units inherited from the Song, which the Ming narrowed and standardized. Ming counties were subdivided into half a dozen or more cantons, cantons into about a dozen townships, and townships into dozens of wards—with yet other units being available for insertion between these levels when population density warranted a finer mesh to the matrix.[52] A ward was small enough—mandated as fifty families in the Yuan, a hundred or so families in the Ming—to conform to the contours of existing villages, or that was the ideal. Onto this spatial configuration, the Ming in 1381 imposed a registration system known as the *lijia* (hundred-and-tithing) system. Ten households made up one *jia* (a "tithing"), and every ten *jia* made up one *li* (a "hundred," to use the medieval English equivalent). Add to this hundred households the ten most prosperous families in the neighborhood, who were required to direct activities and collect taxes, and you had a ward, at least in theory. This state-organized system of communities extended uniformly across the entire realm. Not a household was permitted to evade the system, and in the early Ming very few did.

The state not only tagged every household, but registered every able-bodied adult male and every plot of land for taxation. Households and their labor were recorded in the Yellow Registers, and land in Watertight Registers *(liushui pu)*—signifying that not a drop of land should escape registration. This register listed every field under tax assessment at the ward and township level. From the sketch map at the front of each register summarizing the data, on which plots were drawn like scales on a fish, these booklets came to be known as Fish-Scale Registers.[53]

Just as the Ming aspired to place every person within the administrative matrix, so it aspired to measure every plot of arable land. Accurate land data were "the highest expression of benevolent government," writes Lu Rong in his commonplace book. A just realm depends on equitable taxation, and taxation is equitable only when everyone is taxed at the same standard. "If one appoints the right people to carry out the order to measure land and determine the actual size of holdings, then it is possible to clarify who is paying too much tax and who is paying too little."[54] Surveying fields accurately was not easy. In an administrative handbook he wrote while serving as a magistrate from 1558 to 1560, the famously unbending official Hai Rui (1513–1587) gave precise instructions for surveying taxable land. The first rule was that survey maps must be drawn consistently using an accurate east-west axis determined by sunrise and sunset, not local custom. The second was that they must be accurate, with measurements taken under less than ideal conditions verified later when the sun shone. The third is that they must be done with uniform measures. Hai has included a great deal of other practical advice, such as how to survey hill land from multiple perspectives. Proper surveying was essential, as it "clarifies boundaries and equalizes taxes, thereby relieving the people's poverty and bringing their lawsuits to an end."[55]

The ideals that Hai Rui laid down were often evaded in practice, just as tax obligations had a way of floating away from those who owned the land and falling on the backs of those least able to pay them, especially when the clerks keeping the registers could be bought off. In 1580, the hard-working grand secretary Zhang Juzheng (1525–1582) decided to get to the bottom of the mess. He called for a new set of books for the entire county. Perhaps he was driven less by a sense of justice than by a desire to improve state revenue, but the outcome would be the same. He ordered all county magistrates "to resurvey all land in the realm for tax purposes so that not one inch of land was omitted."[56] Zhang was on the way to getting the information he needed by the next Great Compilation year of 1582, but he died that year and the inch-by-inch administrative web he dreamed of extending across all of Ming space was not completed.

Scale remained the problem. The realm was too big for every place to be under direct surveillance from the center. The imperative of centralization, however, was too strong to permit authority to devolve or local administrative practices to vary. In practice, of course they did, hugely.[57] Yet Ming officials managed nonetheless to work out a compromise between these contrary tendencies, producing as thorough a system of administrative control as a preindustrial state could hope to develop.

3

THE NINE SLOUGHS

RESIDENTS of the small prefectural city of Qiongshan on Hainan Island could see them approaching on that summer's afternoon in 1458. Nine dragons floated above them on multicolored clouds. Hainan is the large subtropical island lying off the south coast, Guangdong's and China's most southern point. Something of a frontier zone even as late as the Ming, it was a place where Chinese from the mainland and the aboriginal peoples living in the mountainous jungles coexisted uneasily. The island was on the front line of typhoon landings, especially during the monsoons that brought China its annual quotient of rainfall every summer.

The squad of dragons swooped down from the clouds and attacked the county magistrate's *yamen,* utterly demolishing the front gate. Was the magistrate being put on notice, or was the real target the Tianshun emperor (r. 1457–1464) who, just the year before, had usurped the throne that had passed to his half-brother when he got himself captured by the Mongols on a hare-brained expedition north of the border? Having delivered this blow to the state, the dragons then turned on a woman, choking her to death and tearing her corpse apart, uneasily echoing the extreme punishment with which the state sentenced the very worst offenders, known as *lingchi* or "death by a thousand cuts." What was her crime, we are made to wonder? The dragons then rose and veered northeast, knocking down the homes of the poor and scattering their meager contents across the townscape. Millions of dragonflies trailed in their wake. Later that summer, an enormous typhoon slammed into the island, uprooting trees and flattening houses.[1]

That one summer the world on Hainan Island was briefly turned upside down. Though not constantly harried by such environmental extremes, Hainan did experience its share of natural disasters, judging from the annual record preserved in the island's 1618 prefectural gazetteer. The dragon attack was but one of the bewilderingly varied disorders of nature visited on the island during the Yuan and Ming. Between the earliest entry (1305) and the last (1618), the people of Hainan Island experienced torrential rains (1305, 1520, 1585), famines (1324, 1434, 1469, 1528, 1572, 1595, 1597, 1608), droughts (1403, 1555, 1618), locusts (1404, 1409), dust clouds (1412), typhoons (1431, 1458, 1542, 1616), earthquakes (1465, 1469, 1523, 1524, 1595, 1605), food shortages (1469, 1572, 1596), fires (1479, 1588), bizarre creatures (a bat-winged feline landed in the Confucian temple in 1482, and a pig birthed an elephant in 1496), epidemics (1489, 1506, 1597), meteors (1498, 1610), snow (1507), multiple births (triplets in 1509), changes in sediment in the water (1511), fierce wind (1515), flood (1520), tsunami (1524), hail (1525, 1540, 1618), petrification (1539), lightning (1582), and cold so intense that animals froze to death (1606). To this litany of nature's anomalies, the gazetteer compiler adds the social effects of these disruptions, such as banditry (1305), mass exodus (1595, 1608), and a rebellion among indigenous islanders (1612).[2]

Hainan Island was considered an outlier—quite literally, lying as it did off the south coast of Guangdong province, but culturally as well. A native of the Yangzi delta who was appointed prefect of the island later in the century told a friend that he acquired twin servants there, one named White Dragon Boy and the other Black Dragon Boy. Their bodies bore no special marks, but they could swim incredible distances under water and preferred raw seafood to cooked food. They were this way because their mother had been impregnated by a dragon, or so he was made to understand.[3] They did things differently on Hainan Island, and the multiple disturbances of nature seemed only to prove the point.

The compiler of the gazetteer meant this list to reflect not backward on some peculiarity of the island, however, but forward onto the record of those appointed to administer it. Did they care for the people when troubles overtook the island, or did they fail? We can use his list for another purpose, not just as a chronology of administrative stress on this one island, but as evidence of environmental stress and climate change during the Yuan and Ming dynasties. Weather is the most basic of life's material conditions, the joker that can transform an ordinary crop into a bumper harvest or usher in the most calamitous disaster by withholding any of the conditions essential for crops to ripen and for farmers to sow and reap them.

Comprehensive lists of weather anomalies and disasters have been included in the Five Phases or Five Elements chapters of the dynastic histories. In the *History of the Yuan* the recorded disasters include floods, unseasonable frosts and snows, hail, thunder, fire, lack of snow when snow should have fallen, ice storms, excessive rainfall, droughts, locusts, famines, epidemics, high winds, insect infestations, landslides and earthquakes, and, of course, dragons. The lists in the *History of the Ming* are slightly less dense with entries, perhaps because, by comparison, the anti-Mongol compilers of the earlier history were particularly keen to attribute as many disasters to their unloved former rulers as they could.[4] In any case, the data in these histories allow us to sketch a chronological profile of the natural and social calamities that shaped these four centuries. The dynasties were not utterly submerged by disasters; there were good years, as we shall see. But the bad years brought occasional shocks that could be deadly.

To these data may be added references scattered through the diaries and commonplace books that people of the time kept. It is unfortunate that the weather diary kept by Zhou Chen (1381–1453), who became famous throughout Jiangnan for bringing order to the fiscal confusion that had plagued the Yangzi delta since the Ming takeover, has not survived. Zhou kept a daily record of weather conditions as a device for verifying the truth of witness statements in law cases. When questioning a witness, he would ask what the weather was that day, check the reply with his weather diary, and then decide whether he was telling the truth. "Recording the wind and rain is indeed public business, not idle jotting," concludes Lu Rong, who preserves Zhou's trick in his commonplace book, *Miscellany from Bean Garden*.[5]

Some also kept records in the hope of discovering patterns that might reveal the logic of disasters yet to come. Thus one gazetteer editor in 1630 could look back to a huge snowstorm that blanketed Shanghai in three and a half meters of snow in January 1445 and declare it to have been a portent of a Japanese pirate attack later that year. So too he linked a rooster that uttered human speech in 1555 to another wave of Japanese piracy that same year.[6] The compiler of the Hainan gazetteer was less ready to find such literal meanings in these disturbances. After all, he observes, "when the mind is rectified and the physical self in harmony, Heaven and Earth will also be so."[7] The culprit was not Heaven but bad administrators. As we will see, however, even the most earnest official sometimes found himself struggling with weather anomalies so extreme that the majority of the stricken remained beyond rescue.

Cold

The Yuan and Ming dynasties belong to a period of climate anomaly historians call the Little Ice Age. Starting about 1270, the earth became a colder planet than it had been during the previous quarter-millennium (known as the Little Climate Optimum or the Medieval Warm Period). The first phase of cold temperatures reached its nadir about 1370, followed by a mild reprieve lasting about a century. Global cooling resumed about 1470, lowering temperatures further and dropping snow on places that had never experienced it. The depth of snow that fell on Florence in 1494 was so great as to inspire the ruling family there to commission a giant snowman from the sculptor Michelangelo. Temperatures became even colder through the sixteenth century, though this trend was occasionally relieved by periods of warming. Temperatures fell again around 1630, reaching their coldest point in the millennium in 1645 and remaining there until 1715.[8]

The Little Ice Age has been reconstructed largely on the basis of data from outside Asia. What is the evidence from China? Climate anomalies can be detected from variation in the thickness of the annular rings in trees, but the depletion of forests during the Yuan and Ming deprives us of long-term tree-ring data. Change in the rate of glacier growth is another widely used indicator, and this we have for China. Radiocarbon data taken from glaciers on the Tibetan Plateau stretching west from Sichuan indicate that glaciers started advancing there in the latter part of the thirteenth century: the Little Ice Age was settling in over Asia at roughly the same time as Europe.[9] Using these and other physical data, meteorologists Zhang Jiacheng and Thomas Crowley have identified the later phase of the Little Ice Age in China after 1450 as particularly cold, the lowest temperatures falling in the middle of the seventeenth century.[10]

This profile is richly confirmed, and made more precise, when we turn to indicators of weather extremes in the dynastic histories and local gazetteers. These published records show that for only one year of the Yuan dynasty, 1316, is there any indication of the weather being warmer than usual. Otherwise, all reports show weather to be normal or colder than normal. We cannot date the poet Zhang Yuniang to better than the reign of Khubilai Khan, but I like to think that she wrote "Singing of Snow" between 1284 and 1294, Khubilai's coldest decade and his last:

The heavens are shrouded,
Red clouds dark and murky, cold wind rising fiercely.
Cold wind rising fiercely,
Shards of shattered gems
Ride the wind, swirl on gusts.

Those beauties ought to despise their own lightness,
Snow-white shapes recklessly tossed against the bed-curtains.
Tossed against the bed-curtains,
They cannot but feel chilled—
But to whom can they speak?[11]

Ming temperatures show more variation, yet cold weather predominated. The dynasty started out cold, recovered to normal temperatures between 1394 and 1438, then turned cold again and stayed that way over the next decade and a half. The winter of 1453 was harsh, with unusually heavy snowfalls settling across the land from Shandong in the northeast of the country to Jiangxi in the center. That April, the minister of personnel reported to the emperor that down on the Yangzi delta, "the number of people who froze to death was high," reaching 1,800 in Changshu county on the south shore of the Yangzi and rising well above that on the north shore.[12] The following spring, bitter cold caused bamboo to freeze and the Yangzi estuary to ice over. The next winter, snow blanketed the entire delta to the depth of a meter. The harbors on Lake Tai froze, forcing all boat traffic to cease. Animals perished in great numbers.[13]

This period of severe cold ended in 1456, giving way to a warm spell of three years. The weather then fluctuated wildly for the next sixty-six years, veering from cold to warm and back again, sometimes between one year and the next, though cold spells always outlasted warm ones. In the winter of 1477, the cold was fierce enough to freeze the canals lacing the Yangzi delta to a depth of several feet, shutting down the region's transport network for several months. Further cold bouts followed over the next four decades. This volatility ended in 1536 when, for the next thirty-five years, the Ming experienced its only extended warm spell. But the weather turned cold again in 1577, and the following winter, lakes on the Yangzi delta froze and winds blew the ice into mounds ten meters high.

Other than two warm years in 1589 and 1590, the late Ming stayed cold. When the missionary Matteo Ricci headed south on the Grand Canal after his first visit to Beijing in the winter of 1597–1598, he discovered that "once winter sets in, all the rivers in northern China are frozen over so hard that navigation on them is impossible, and a wagon may pass over them."[14] (This in no longer true.) Temperatures remained cold through to the end of the dynasty, sinking to an extended period of unprecedented cold from 1629 to 1643.

It is tempting to align the political fortunes of the Yuan and Ming dynasties with their temperature profiles: Khubilai Khan moving his regime south to Beijing just around the onset of the Little Ice Age; his dynasty collapsing in 1368, at the peak of the first phase of the Little Ice Age; the Ming in turn collapsing in 1644 at the end of the most extended bout of severe cold weather on record for these four centuries. Temperatures are not the only factor explaining these larger events, but they must be part of any explanation.

This feature of history is illustrated by a source that China historians have not considered as an indicator of weather: paintings. We are familiar with European snow paintings from the Little Ice Age by artists such as Pieter Brueghel the Elder (ca. 1525–1569), Hendrick Avercamp (1585–1634), and Thomas Heeremanns (ca. 1640–1697). But while the canals of the Netherlands and Belgium were icing over in the sixteenth and seventeenth centuries—and artists were painting these scenes for their novelty—so too were the canals in north China. Chinese painters were less inclined to document their immediate surroundings than were European artists of the period, yet it might be worth looking at snowscapes and seeing what we get. We tend not to think of China as a snow country, but the popularity of snow paintings at certain times in the Ming (too few survive from the Yuan to make any comparison) suggests that artists were representing more than the idea of snow.

The most prolific Ming painter of snow scenes turns out to be the first: the widely influential court painter Dai Jin (1388–1462). Aligning his dated work with the weather, we find that Dai painted snow scenes during the first period of colder weather during the Ming, from 1439 to 1455 (Fig. 3).[15]

Fig. 3 *Returning Home through the Snow* by Dai Jin (1388–1462). Dai painted this evocative wintry scene about 1455, the final year of the Ming dynasty's first prolonged bout of severe cold weather, which started in 1439. Metropolitan Museum of Art, New York.

The next cluster of snow paintings appears in the early years of the sixteenth century, when Tang Yin (1470–1524) and Zhou Chen (d. after 1535) were loosening the academic style of their predecessors and infusing painting with a new creativity. The paintings coincide with the cold bout of 1504–1509.[16] Then comes a tight cluster between 1528 and 1532, which was the period when painting was dominated by the century's most influential artist, Wen Zhengming (1470–1559). Wen was the most prolific painter of snow scenes among his cohort, and indeed is credited with making snowscapes popular. Though this was not a major cold phase, cold years punctuated the period (in 1518, 1519, 1523, and 1529). Wen's *Heavy Snow in the Mountain Passes,* painted about 1532 (Fig. 4), is one of the richest portraits of a snow-laden landscape in the entire oeuvre of Ming painting.[17]

The next group of snow paintings belongs to the latter half of the long Wanli era (1573–1620). The leading figure in this group of snow painters is Dong Qichang (1555–1636), whose art practice and theory redefined the conventions of taste in Chinese art. His *Eight Views of Yan* belongs right in the middle of the cold spell of 1595–1598. Dong once declared that he disdained snow scenes. "As a rule I never paint snow," he is reported to have said. "I let wintry scenery serve in its place." The painter who recorded this statement complains about the lack of snow in Dong's work, declaring that his snowless winter scenes look just like autumn scenes: "winter scenery by definition."[18] Dong's artist-friends were less reserved when it came to snow. Zhao Zuo, a close friend, painted at least two snowscapes in the dynasty's penultimate cold period of 1616–1620.[19]

The year in which Zhao painted *Piled Snow on Cold Cliffs,* 1616, is a year for which we happen to have the daily record of the art collector and connoisseur Li Rihua (1565–1635), whose diary is the subject of Chapter 8. This was the harshest winter on the Yangzi delta in many years, though it started late that year. Only on January 3 did the west wind begin to blow. Nine days later the temperature fell. Li's entry for January 12 consists of two words: *han shen,* "cold in the extreme." The first snow fell on January 29. The weather turned warmer, but on February 4 the snow returned. "Fine snow in the evening," he writes on February 18; the next day, "heavy snow"; the day after that, "it is still snowing and has accumulated to a depth of four or five inches. This sort of weather has not been seen for the past six or seven years."[20] Li planned to go to Hangzhou on March 19 but was stopped by snow. Perhaps we have that winter's snowfall to thank for inspiring Zhao Zuo to paint snow.

Fig. 4 *Heavy Snow in the Mountain Passes* by Wen Zhengming (1470–1559). Wen was fond of painting snowscapes. He did this one sometime between 1528 and 1532; 1529 in particular was a cold winter. National Palace Museum, Taiwan, Republic of China.

The final burst of snow paintings belongs to the final eight years of the dynasty, between 1636 and 1643. This, as we have noted, was when the Ming was going through its coldest phase. The most prolific of these painters was the highly original Zhang Hong (b. 1577), whose style shows influences from European art.[21] Many of Zhang Hong's contemporaries, including Wang Shimin (1592–1680) and Lan Meng (b. 1614), were also painting pictures of the snow.

Drought

If the Yuan and Ming dynasties were more often cold than warm, the dynastic histories also reveal they were more often dry than wet.[22] The Yuan opened in dry weather that lasted for four decades. The weather turned wet early in the fourteenth century (1308–1325), fluctuated for a time between extremes of wet and dry, then entered a second dry phase extending from the end of the Yuan through the beginning of the Ming (1352–1374). The first quarter of the fifteenth century was wet, then drought struck in 1426. Except for a few wet interludes in the 1450s and the 1470s, the drought continued past the end of the century. Precipitation returned to normal in 1504, with occasional wet periods, until 1544, when drought seized the realm. For the final century of the dynasty right down to 1644, the Ming was abnormally dry. The dryness peaked three times, in 1544–1546, 1585–1589, and 1614–1619. The last of these three droughts parched the fields so thoroughly that the *History of the Ming* reports the landscape in 1615 looking burnt.[23] The dynasty ended in seven years of devastating drought.[24]

When drought struck, officials appealed to the dragons to bring rain. During a drought in 1563, the prefect of Hangzhou dispatched a Daoist priest to a mountain in the prefecture to capture a dragon to whom sacrifices might be made at the state altar in Hangzhou. When the priest reached the Dragon Pool on this particular mountain, all he caught were four frogs and one bullfrog. He put the bullfrog in a pot to take back to the prefect. The journey proved to be troublesome, and he complained aloud to his retinue, "So much tax and labor has been spent on this frog that if it does not respond to my prayers, I will cook and eat it when I get back." Immediately there came a thunderstorm that soaked the priest. When he looked in the pot, the bullfrog had disappeared. He hurried back to the Dragon Pool to offer prayers of repentance for his indiscrete remark. He caught another bullfrog, took it back to Hangzhou unharmed, and prayed before it, causing rain to fall. He then took the bullfrog back to the mountain in the same pot. When he opened the lid to release it back into Dragon Pool, he discovered that it too had disappeared.[25] The bullfrog was deemed to be an avatar of the Dragon Lord himself.

Floods

Floods are complex events. The normal rains that fall in the wake of a drought can start a flood just as much as excessive rainfall can. And whether a river floods or not depends at least as much on the investment that the state makes in maintaining dikes and dredging river beds as it does on the amount of rain that falls in any one year.

The compilers of the *History of the Yuan* were particularly attentive to floods. They record more in that one century than the compilers of the *History of the Ming* do for the three centuries that followed. While there may well have been more floods during the Yuan century, there was certainly more enthusiasm for reporting them. They began in the 1280s, but not until the year after Khubilai's death did the flood gates really open. The Yangzi River flooded through the summer of 1295, turning what began as a series of separate local disasters into a national calamity. Flood struck the upper reaches of the river the following summer, and then the Yellow River broke its banks in several places that fall. Through the winter, spring, and summer following, in locality after locality, in the simple phrasing of the dynastic historians, "the fields and houses disappeared under the waves."[26] All of China, it seemed, was under water.

A longer phase of flooding began in 1301, and floods recurred on an almost annual basis thereafter. Khubilai's descendants were thus fated to rule—or misrule—a flooded realm. The floods through the years from 1319 to 1332 were particularly intense. Rivers overflowed their banks, lakes inundated the surrounding countryside, and tidal waves surged inland along the coast. Walls and dikes built to channel the eastward flow of water across the Chinese landscape were washed away. When a coastal seawall burst in April 1328, the court ordered a group of Tibetan monks to cast 216 statues of the Buddha and pray for divine intervention, to no avail: a tidal wave submerged the statue-studded coast the following month. At that point, the court launched the more expensive response of recruiting local military and civilian labor to build ten miles of new stone embankments.[27] In 1346 the flooding relented, returning significantly only twice through the dynasty's last two decades. The respite came too late to convince the people of the Yuan that the Mongols had not lost Heaven's mandate to rule.

The Ming dynasty experienced only sporadic flooding until the 1410s. There were particularly severe floods in 1411 and 1416, coinciding with the reconstruction of the Grand Canal. Flooding returned in the mid-1440s and again in the mid-1450s. Not until 1537 was the Ming struck by truly massive floods. Serious flooding occurred in 1569 and again in 1586, and lesser floods followed these. Mercifully, with the exception of 1642, the last three decades of the dynasty were flood-free.

Locusts

China has always been vulnerable to crop-devouring insects such as locusts. The intensity with which the people of the Yuan and Ming practiced agriculture only increased this vulnerability, and not just on the North China Plain but throughout the Yangzi valley as well. The *History of the Yuan* records serious locust infestations almost every year, as persistent as floods. In many years, the two coincided, notably during the terrible floods of 1295–1297 and the fierce plague of locusts in 1328–1330. During the latter infestation, locusts eradicated crops to a degree that would not be seen again for a century. There was some respite through the next two decades, then they struck again in 1349, and did so intermittently through the last five years of the dynasty.

Other than a four-year bout of locust attacks in the 1370s, the early Ming was remarkably free of voracious insects until 1434–1448. The locust plague of 1441 was particularly devastating. They returned in the late 1450s (1456 being an especially bad year) and the early 1490s, but otherwise did not cause the destruction they did in the Yuan. Locusts struck seriously only three times in the sixteenth century (1524, 1569, 1587). It is through the first half of the seventeenth (1609, 1615–1619, 1625, 1635, 1637–1641) that the insects returned with increasing energy. In those last five summers, locusts destroyed crops on a scale that the people of the Ming had not seen since they were the people of the Yuan.

Locust infestations tended to occur when rain brought a prolonged drought to an end. The onset of the most virulent locust attack during the Yuan dynasty, in 1328, came three years into a severe bout of dry weather. Similarly, the first sustained infestation of the Ming period, in the early 1440s, started in the fifth year of a long drought; and the locust plague of the early 1490s broke out exactly in the middle of the next protracted spell of dry weather (1482–1503). The arrival of rain after several years of drought tends to stimulate the reproduction of crop-devouring insects, and this indeed is what seems to have happened in the Ming. Every major infestation in the dynasty's last century arrived in the wake of a drought.

Earthquakes

The topography of eastern Eurasia is the outcome of the convergence of several microcontinents, and earthquakes are its most vivid manifestation. China is a jigsaw puzzle of tectonic plates rubbing against one another. The three major seismic fault lines east of the Tibetan Plateau run roughly north-south: in Shanxi down the Fen River valley to where it converges with the Yellow River at the point it turns east; off the coast of Fujian; and up from Yunnan through the Sichuan Basin. All three zones were active during the Yuan and Ming.

The Yuan experienced a spate of earthquake activity in 1290–91, but nothing like the quakes that rocked the Fen River valley starting on September 13, 1303. The people in the county town of Gaoping, which sits midway up the river, were awakened from their beds shortly after midnight by a terrific wind that blew in from the northwest. The awakening was a stroke of good fortune. When the earthquake (estimated at 5.5 on the Richter scale) struck a few hours later, most people were up and outside their houses, the majority of which collapsed. Riding the quake, they reported, felt like rowing a boat across a river.

The Gaoping earthquake proved to be modest compared to the second quake that struck four nights later 50 kilometers up the Fen River in Zhaocheng. The force of the Zhaocheng quake (8 on the Richter scale) was enough to flatten anything that the Gaoping quake had left standing. Buildings as far away as the Yellow River were reduced to rubble, with tremors felt well beyond that. Between a quarter and half a million people were crushed in the first wave. Hundreds of thousands were left injured, and hundreds of thousands of buildings were destroyed. Seismic aftershocks kept the region unsettled for another two years, and the drought that descended in the wake of the earthquake refused to release its grip on the province for another year after that.[28] Earthquakes continue through the rest of the dynasty, recurring almost annually between 1338 and 1352.

The Ming was shaken in its early years, then in the 1440s, the 1480s, and 1505–1528. The last of these phases gave rise to a curious instance of dragon spotting in Yunnan. In his commonplace book, *Notes from the Last Two Years of the Zhengde Reign (Gengsi bian),* Lu Can (1494–1551) records an odd report of a chalk-white dragon appearing at midnight in a garden in the early years of the Zhengde emperor's reign. The garden belonged to a local degree-holder named Wang Cheng. The place was called Dragon Guard, and it sat within a special administrative zone where Yunnan borders Burma. The dragon's scales were sharp enough to cut the hand of anyone who touched them, but the creature lolled about and showed no inclination to depart. Wang grew anxious lest the crowds of curious who came to see it over the next few days grow out of control, so he resorted to an old dragon-expelling technique. He smeared it with dog's blood. The device worked, and the dragon disappeared. Lu Can explains that the Wangs noticed the dragon because they were sleeping in the garden in makeshift sheds, as they had been for half a year—waiting out the aftershocks of a major earthquake.[29]

The first Ming earthquake to match the great Zhaocheng quake of 1303 struck in the same region on January 23, 1556, this time up the Wei River instead of the Fen River. Estimated at 8 on the Richter scale, it scythed a 250-kilometer path of destruction down the Yellow River valley and up the Fen, leveling city walls, government buildings, and homes. At the epicenter on the Wei, the entire housing stock was demolished, half the residents were killed, springs changed location, and rivers flowed in new directions. The provinces of Shaanxi and Shanxi continued to shake for a month, with shocks being felt as far northwest as Gansu, as far east as Shandong, and as far south as the Yangzi. The official death count was 830,000. The actual total probably exceeded a million.[30]

The last great Ming earthquake, on December 29, 1604, struck not in the seismic hot zone up the Yellow River but down off the southeast coast. Although south China lies well west of the faultline where the Philippine and Pacific tectonic plates meet, it is close enough to suffer from its earthquakes. "The ground is constantly moving in Fujian and Guangdong," according to the Fujianese writer Xie Zhaozhe (1567–1624). "One theory has it that, this being a coastal area, there is a lot of water on which the land bobs." Xie was unhappy with this theory, pointing out that Shanxi province had basically no water at all but experienced such violent earthquakes that "the earth splits open to a depth of a dozen meters and the homes of the unfortunate disappear into these cracks, after which they close without even leaving a line to show where to dig. In any case they are so deep that there is no reaching them."[31]

In 1604 the epicenter was only 30 kilometers off the Fujian coast. The quake devastated the two maritime trade centers of the southeast, Quanzhou and Zhangzhou. In Moon Harbor, Zhangzhou's seaport, most of the buildings were destroyed, though the death toll was not what it would have been had Moon Harbor been standing directly on the fault. The earthquake reverberated up the coast as far as Shanghai and even shook the inland provinces of Guangxi and Huguang.[32] Nothing would match it until the great Tianshui earthquake of 1654. Even so, the plates on which the Ming realm floated continued to move, causing major earthquakes for thirty-two of the final forty years of the dynasty. The greatest seismic activity was in the last four years, reminding everyone, if they needed reminding, that all was not well.

Like earthquakes, volcanoes arise from plate tectonics. Unlike earthquakes, however, they occur only at plate boundaries. No plate boundary runs under China, but that does not mean that the Yuan and Ming were free of volcanic effects. For just as earthquakes along the Philippine fault could be felt on the mainland, volcanoes in this zone spewed aerosol debris that was blown over China. Volcanic clouds decrease the light and heat that reaches the earth's surface, often for months after the initial eruption, forcing rapid climate anomalies that can diminish harvests and induce famine.

Given the volcanic activity in Japan, Ryukyu (Okinawa), and the Philippines during the Yuan and Ming, it would be surprising if none of their aerosol spumes cast a westward shadow. And indeed there are some suggestive coincidences: between the eruption of Azama (Japan) in 1331 and the cold phase of 1330–1332; the eruption of Iraya (Bataan) in 1464 and the cold phase of 1464–1465; the eruptions of Iwaki and Asama (Japan) in 1597–1598 and the drought and famine of 1598–1601; and the eruption of Iriga (Luzon) in 1628 just prior to the onset of the cool phase of 1629–1643.[33] Were these particular eruptions powerful enough to block the sunlight warming Ming fields and ripening its grain?

Epidemics

Epidemics struck the people of the Yuan and Ming many times, though with particular severity in four periods: over the last fifteen years of the

Yuan (1344–1345, 1356–1360, 1362); between 1407 and 1411, with 1411 being the worst outbreak of the fifteenth century; during the widespread catastrophes of 1587–1588; and over the last six years of the Ming (1639–1641, 1643–1644). The four decades from 1506 to 1546 also had more than their fair share of sickness. The last three waves of epidemics in the Ming coincide with the three worst drought-induced famines of the dynasty's last century: 1544–1546, 1587–1588, and 1639–1641.

The forensics for these disease events are difficult to perform. All we have to go on are descriptions of symptoms recorded by people who for the most part had no medical knowledge and whose knowledge in any case does not overlap with our own. Dysentery, typhoid fever, smallpox, and plague are the likeliest candidates. Historians have been particularly susceptible to suggestions that the waves of sickness in the late Yuan and the late Ming may have been bubonic plague, a bacterial infection spread by bites from infected rat fleas. The attraction of this diagnosis has much to do with Europe's dramatic experience with the plague in the fourteenth century, and not just because of the horrifying scale of the epidemic. It was the Mongols, after all, who infected the first Europeans—the Italians against whom they were laying siege on the north coast of the Black Sea—who transferred the disease back to Constantinople and Italy in 1347. The originators of the plague were warriors of the Kipchak Khanate, otherwise known as the Golden Horde, the Mongols who broke off from Khubilai when he founded the Yuan, forming an independent realm at the western end of Chinggis's original empire.

The Singaporean epidemiologist Wu Lien-Teh, dubbed China's plague fighter for organizing the program to control the virulent outbreak of pneumonic plague in Manchuria in 1911, was the first medical scholar to propose that the late-Yuan epidemic was plague, albeit it on limited evidence. That the plague could have spread so rapidly across the Eurasian continent appealed to the world historian William McNeill, who adopted this assumption as the backbone of his 1976 masterwork on the global history of epidemics, *Plagues and Peoples*. The presence in modern times of a large plague reservoir on the Mongolian steppe, where the Mongolian gerbil hosts the Asiatic flea, lent the idea additional weight.

The demographic historian Ole Benedictow has now called this assumption into question. "The Black Death has made a profound impression on the scholars who have studied it," he observes, to the degree that "their sense of the extraordinary and exotic" has overwhelmed their critical faculties. Problems of origin are better approached not by casting about for the most exaggerated hypothesis, he argues, but by working from "the principle of proximate origin." That is, "the shorter the distance to be covered, the fewer obstacles to dissemination."[34] Benedictow argues persuasively that the obstacles were too great to allow the plague to travel from the Yuan realm in 1344 to shores of the Black Sea in 1346. Two years is too short a time for the bacteria to move such a great distance. The Kipchak Mongols made such a quick transfer even more unlikely by cutting off the caravan trade between China and Europe in 1343, the year before the outbreak of epidemic sickness in the Yuan.

The obstacles were space and time. Plague could cover forty kilometers a day in the fourteenth century when carried by rats or humans onboard ship. Overland, however, the rate of travel fell to less than two kilometers a day. Given that bubonic plague takes only three to five days to incubate, that the stricken are infectious for less than a day, and that 80 percent die within three to five days, it is hard to imagine the infection covering even two kilometers, and doing so consistently enough to pass the bacteria down the full length of the Silk Road. Uzbekistan or some other location within the Golden Horde is the likelier site of origin. Finally, there is the argument from climate. Fleas have a high natural mortality and have to reproduce themselves constantly to sustain their population; cold weather inhibits flea procreation.[35] The years 1344–1353 being among the coldest in the Yuan, conditions for the long-distance transfer of plague fleas were about as unpropitious as they could be.

The assumption that the epidemics at the end of the Yuan were plague has been applied also to the sickness at the end of the Ming. The demographic historian Cao Shuji has proposed that both the 1587 and 1639 outbreaks were plague. He traces the former to the reopening of border trade with the Mongols in 1571. He attributes the outbreak as well to the disturbance of the habitat of the Mongolian gerbil, a potential plague carrier, by Chinese farmers migrating out onto the Mongolian steppe. Cao suggests that the jump from rats to humans happened about 1580, the year an epidemic started in Shanxi province. The Ming shut down the horse markets as soon as this happened, but it was too late. The sickness reached Beijing in May 1582. It went dormant for several years, then reerupted there in June 1587, and again exactly one year later.[36]

A Fujian writer who was in Beijing at the turn of the seventeenth century blamed the outbreak on the deplorable living conditions in Beijing.

"The houses in the capital are so closely crowded together that there is no open space, and the markets teem with excrement and filth," he complained. "People from all directions live together in disorderly confusion, and there are many flies and gnats. Whenever it gets hot, it becomes almost unbearable." Cooling rain was not always the solution, however. "A little steady rain has only to fall and there is trouble from flooding, so that malarial fevers, dysentery, and epidemics follow one on the other without end."[37] Whether the same bacteria were the cause, the description is eerily similar to the picture Edward III drew for the mayor of London during the plague in 1349: "The streets and lanes through which people had to pass were foul with human faeces and the air of the city was poisoned to the great danger of men passing." In another letter to the mayor in 1361, the king complained about "putrid blood running down the streets," noting that "the air in the city is very much corrupted and infested, hence abominable and most filthy stench proceeds, sickness and many other evils have happened to such as have abode in the said city, or had resorted to it; and great dangers are feared to fall out for the time to come."[38]

Emperor Wanli did not have the same opportunities as an English king to go out into the streets of his capital, but his chief grand secretary, Shen Shixing, reported the outbreak of the 1587 epidemic in the city to him on June 11. "The weather is hot and dry and rain brings relief so rarely," he told Wanli, so that "wherever the vapors of pestilence are detected, epidemic flourishes." Shen reminded the emperor that the Hongwu and Jiajing emperors supported public pharmacies in similar circumstances and appealed to him to do the same. He suggested that Wanli "order the Ministry of Rites to instruct the Court of Medicine to increase its distribution of medicines and despatch carefully selected medical officers to several locations inside and outside the city walls of the capital to give medicine to the ill, so as to make manifest the dynasty's concern for the lives of its subjects."[39] The emperor followed Shen's advice, though he soon found himself up against his ancestors' fixed limits on cash disbursements to the destitute. No Ming emperor was supposed to revoke directly what his ancestors had laid down. Wanli got around them by invoking a precedent from his grandfather's time authorizing payments to patients to cover medicine costs. This sort of intervention had only a limited impact on mortality. Demographic historian Cao Shuji estimates on the high side, proposing death rates in north China of between 40 and 50 percent. This would have reduced the population of the three northern provinces of Shanxi, North Zhili, and Henan from 25.6 million in 1580 to 12.8 million in 1588.

The Yangzi valley was also struck by virulent epidemics in the same year. This outbreak may have come down the Grand Canal from the north, as reports of the sickness tended to appear along the Grand Canal corridor. An equally good argument can be made, however, for this being a deadly cocktail of endemic water-borne diseases such as typhoid fever and dysentery that erupted due to a general weakening of health brought on by the 1587–1588 famine.[40]

Children are especially vulnerable to infectious diseases (Fig. 5). But the epidemic in the Yangzi Valley has left a curious bit of evidence in the form of a poem written in 1588 by a widow named Madam Chen:

> A year of flood and drought, a year of great distress,
> All shut their doors to the sickness passing from house to house;
> Do not spurn the orphans who have nothing to eat but gruel,
> Instead give thanks that Heaven keeps them among the living.[41]

Madam Chen's poem shows that poeple had a clear sense of the danger of being in contact with the infected, and that they quarantined themselves for their own safety. Her reference to orphans, however, is intriguing for it suggests that some children were surviving the sickness that was killing their parents. This would be unusual in an epidemic, given that the very young and the very old are the first victims of infectious disease. If disease rather than malnutrition was indeed the main cause of death, then some children developed resistance to whatever it was that was killing their parents.

Famine

Hunger was not a constant in this period, but it did return regularly, particularly in the Yuan. Between that dynasty's first phase of dearth (1268–1272) and the last (1357–1359), people experienced a major famine on average once every two years. The worst decade was the 1320s. The political chaos during this decade, when the dynasty went through five emperors, each one younger and less powerful than his predecessor, must have contributed to the failure of the regime to stem these disasters. We could just as easily turn this equation around, though, and wonder

Fig. 5 Guanyin, the Chinese manifestation of the Buddhist god of mercy Avalokitesvara, intervening to rescue a child being carried off by a demon of pestilence. This seventeenth-century painting postdates the terrible epidemics of the early 1640s. Royal Ontario Museum, Toronto.

whether the train of famines created that sense of instability. Court conflict dwindled in the 1340s, but by then the weather was doing most of the heavy lifting.

The worst hunger years of the Yuan were still remembered centuries later. Haiyan county, the site of Chen Mountain on the Yangzi delta where the Dragon Lord and his son relieved a drought in 1293, was devastated by famine in 1539. Haiyan elders on that occasion were moved to recall a communal memory of a famine in 1305 as the worst famine ever to hit the county. In that year, one elder recalled, "Haiyan county experienced the extreme of famine. People ate each other. It has been over two centuries since we have suffered a harvest failure on this scale."[42] Two centuries was not long enough to have forgotten.

There were short-term crop failures during the first half-century of the Ming, but the first truly severe famine occurred in 1434–1435. During the century from the 1430s to the 1530s, the country experienced a major subsistence crisis every few years. The 1450s was a bad decade, for famine as for every other sort of environmental crisis, but there were also prolonged periods of famine from 1465 to 1473, from 1477 to 1487, and from 1501 to 1519 (with a brief respite in 1515). These bouts of dearth caused distress not just to the starving but to capital officials as well. "The famished, one fears, will use famine as an excuse to rebel," as one minister of war phrased the threat in a messsage to the emperor in 1492. For this reason, local officials should be reminded to act quickly at the first sign of dearth. So too the army should put the troops in a state of readiness to "prepare against the unforeseen." The emperor agreed.[43] Only after 1519 did famine subside.

Drought struck with a vengeance in 1544—an El Niño year—and famine followed in its wake the following year. The 1545 famine was massive. In Zhejiang province, "the lakes dried up completely and became reddened earth." The price of grain skyrocketed. Anyone lucky enough to buy a pint of rice was at risk of being murdered for it as he made his way home in the falling darkness. Beggars died in droves. The government opened its granaries, but supplies were unequal to the need and too slow in delivery. Many of the famished starved to death on the road before they could get to granaries, or died as they waited for grain to be disbursed.[44]

The next severe famine was in 1587–1588.[45] On August 12, 1587, a vice-minister of the Ministry of Revenue reported that people north of the Yellow River were reduced to eating grass and wild plants, and that in southwest Shaanxi their diet consisted of nothing but sand.[46] Drought continued through the spring and summer of 1588, plunging regions that had escaped the famine in 1587 into hunger as well. On April 30, the grand coordinator the court despatched to Guangxi province reported the same crisis in the southwest. Aid was essential to stop the downward spiral, he stressed. A follow-up from a provincial official three weeks later reported that the famine had reached disastrous proportions: "The people are eating each other and the corpses of the famished are scattered about untended. Throughout the cities and countryside are scenes that even a truly skilled painter, were he here, would not be able to paint." At the same time, reported a capital official in Nanjing, people north of the Yangzi "are starving and eating each other," while south of the Yangzi "the price of rice has soared." As he saw it, the onus was squarely on the officials. "Of what use can it possibly be," he asks, "to appoint officials who give the people neither silver nor grain?"[47]

The final wave of famine of the Ming dynasty started in 1632, escalated to vast proportions in 1639, and remained severe for two more years. Neither the Yuan nor the Ming had previously suffered a disaster on this scale. It is the subject of the final chapter.

The Nine Sloughs

The disasters of the Yuan and Ming came in waves. Most waves rarely lasted beyond a season, but sometimes they stretched over two or three years. A seasonal failure was a misfortune people could weather, but a multi-year disaster was something altogether different. To give these extended periods of disaster an identity, I have resuscitated the archaic term "slough" (rhymes with "bough"). A slough is a ditch or low place that collects muck and mires travelers. Since 1678, when John Bunyan wrote of the Slough of Despond in *The Pilgrim's Progress,* it has been used to describe the condition of being bogged down by difficulties. For Bunyan, slough was a metaphor; for me it is closer to a literal description of what life was like during the worst years.

On the basis of the data collected for this chapter, I have identified nine sloughs, three in the Yuan and six in the Ming, and have named them according to the reign era in which they occurred (see appendix).

Khubilai Khan narrowly escaped his dynasty's first major downturn. The Yuanzhen Slough (1295–1297) started the year his successor inherited the throne from Khubilai. Temür may have been unable to match the vigor of his grandfather's rule, but it was his singular misfortune to succeed just as the weather turned bad. Desperate to change his fate, in his third year, 1297, he altered his reign title from Yuanzhen to Dade in the hope of ending the downturn.

The Taiding Slough (1324–1330) struck the dynasty as Khubilai's fifth successor came to the throne—the fifth in three decades. The political instability of the regime was worsened by environmental deterioration, which contributed in turn to the inability of the Mongols to respond effectively. I have dated the end of the Taiding Slough to 1330, but in many ways it did not really end. The wave of trouble simply paused for two years, then resumed in 1333 with an annual series of famines.

The massive drought and flood that initiated the Zhizheng Slough (1342–1345) was followed by floods and major epidemics through 1344 and 1345. It was under these conditions that the future founder of the Ming dynasty came of age. Zhu Yuanzhang's official biography testifies to the role that the Zhizheng Slough played in his early formation by opening his life story with this sentence: "The year 1344 was a time of drought, locusts, great famine, and epidemics."[48] Zhu was sixteen at the time, and these conditions launched him on the road to rebellion. The Yuan dynasty survived for another two decades, its collapse delayed by the civil war among the challengers. By the time Zhu enthroned himself in 1368, the Mongols were irrelevant to the outcome.

The Ming dynasty suffered only intermittent difficulties until 1433, when the weather turned cold, bringing famine, epidemic, and locusts for the next two years. The cold returned in 1437 along with recurring floods through 1448, plunging the realm into a series of famines between 1438 and 1445. These intermittent disasters built relentlessly toward a half-decade of full-blown ecological crisis in 1450, culminating in 1455 in one of the worst years of the century.[49] The Jingtai Slough (1450–1455) coincides perfectly with the reign of the Jingtai emperor, who came to the throne in its first year to replace his half-brother, taken hostage by the Mongols. His half-brother returned and seized the throne in 1456, just as the Jingtai Slough was coming to an end. The Jingtai emperor could not have faced a more miserable set of conditions.

We already know from dragon sightings that the Zhengde era (1506–1521) was a troubled reign. The Zhengde Slough (1516–1519) sealed his reputation then and ever afterward for incompetence and indifference.

The next concentration of disasters occurred a quarter of a century later. The Jiajing Slough (1544–1546) cannot so conspicuously be linked to political crisis. It was simply three years of epidemics and famines on a national scale.

The Wanli emperor enjoyed the longest reign of any Ming emperor—forty-eight years. His longevity rewarded him, however, with two sloughs. We have already noted the severe famine and virulent epidemic that defined Wanli Slough I (1586–1588). Wanli Slough II (1615–1617) was less devastating only by comparison, there being no epidemics, but it was difficult nonetheless.

The final downturn of the dynasty was the spectacular Chongzhen Slough (1637–1643), the subject of this book's final chapter. I have dated this period to the seven years of drought that descended in 1637, but just as easily I could have pushed the onset back to Chongzhen's second year as emperor, 1629, when temperatures fell, setting the stage for the disasters that made his era almost impossible to govern. The slough eased up a few months before the dynasty collapsed, but too late to bring relief.

Dealing with Disaster

The good order of the world was understood to depend on the balance among the Three Powers of Heaven, Earth, and Humankind. Everyone from the emperor to the lowliest person in the realm understood that bad weather meant more than bad weather. It signaled that the Three Powers were out of harmony. There were two ways to interpret this disharmony, one of which endowed Humankind with an active role in its relationship with Heaven and Earth, the other, a passive role.

The active interpretation held that anomalies in Heaven and Earth were due to bad behavior by Humankind. Correct that behavior and the disturbances in nature would cease. This "moral meteorology" placed the emperor in a delicate position, for as Heaven's son he was the human closest to Heaven and should therefore have the greatest influence, for good or ill, with the cardinal Power.[50] This logic meant that every time the weather turned bad, someone could impugn the quality of his rule. A slough thus placed a heavy burden on an emperor. The only thing a censorious courtier had to do to put his emperor on notice was to recite bad weather reports, as one so dramatically did for the Zhengde emperor. The realm was beset with "earth-yellow winds and black fogs, premature spring and winter thunder, earthquakes and dessicated springs, sand in the wind and raining earth," he declared, which meant that "Heaven is transforming above, the Earth transforming below, and Humankind transforming between them." The emperor had better do something about it.[51]

His predecessor, the Hongzhi emperor, was fortunate in not having to rule during a slough. Still, his reign was dogged with bad weather, a situation made worse by the puzzling frequency of dragon manifestations.

Drought was the chief problem. It settled over much of the realm in 1482, six years before his inaugural year, and continued with little break for over the next two decades. This was also a cold period. The fifth year of his reign, 1492, opened under depressed environmental conditions. A general drought was compounded by severe flooding along the rivers of north China. In the opening months of that year, memorials were coming in from all across north China reporting that floods and colder temperatures had caused the autumn harvest to fail. Hongzhi realized he was going to have to write off unpaid taxes from 1491. Holding provinces and counties hostage to past failures would only make the delivery of current-year taxes more unlikely. In three separate tax amnesties, he excused payment of three and a half million liters of wheat.

More troubling to the emperor than the tangible financial cost of disaster was the intangible moral cost. The perception of this cost rose when the court astronomers sent in reports in March that Heaven too was disturbed: an azure comet plummeting southward trailing three small stars in its wake; the moon edging into the wrong constellation; the Wood Star approaching the Altar Star. Then Earth chimed in with an earthquake that produced tumultuous thunder in the far northwest. These were bad omens.

Hongzhi could have prostrated himself before Heaven and begged it to relent, but instead he sought his advisors' advice. What could he, as Heaven's son, do to appease Heaven and Earth? A censor sent to inspect the floods sent in a four-point proposal on March 7 to reallocate the tribute grain coming from the south. The following day another censor proposed a year's moratorium on holiday festivities to cut costs. A bureau secretary suggested that Hongzhi remind his field administrators to be more zealous in the discharge of their duties and order his judges to ease up on extreme punishments. In the language of the officials, these acts should convince Heaven "to transform its disasters into blessings." By his moral example, the emperor would mystically transform the relationship among the Three Powers. But it was not easy. That night the moon edged its way into another constellation where it shouldn't have been. The azure comet returned the night after. It would take time to turn an imperial ship this large around.[52]

Most people could not hope to move Heaven as the emperor could. They tended to adopt a passive interpretation, which was that the disharmony among the Three Powers was periodic. Heaven and Earth regularly wobbled off their pivot. All Humankind could do was put up with this until the period of trouble passed. There was really only one technique available to ordinary humans to deal with the periodicity of mayhem, and that was divination. If you couldn't transform Heaven's disasters, you could at least anticipate their arrival and prepare yourself for them.

The burgeoning publishing industry of the late Ming responded to the demand for divinatory techniques by publishing calendars that listed bad days and divination handbooks that allowed you to peer into the future. Fragments of one such local calendar can be found in the 1574 gazetteer of Cili county in Huguang. The Cili divination calendar starts with the first eight days of the first lunar month, each of which governs the fortunes of chickens, dogs, pigs, goats, oxen, horses, people, and grain in turn. If the weather is sunny on the day for chickens, the first day, then they will flourish; if it is cloudy, they will meet with disaster. The same rule applies to dogs on the second day, pigs on the third, and so forth. Clearly the two most important days are the seventh and eighth, the weather on these days determining how people and crops will do through the coming year.

The Cili prognostications also tagged particular days through the year, often illustrating them with a local adage. For the third day of the third month, for example, if you hear the sound of frogs croaking before noon, crops at higher elevation will ripen, whereas if you hear the croaking after noon, crops at lower elevation will ripen. The last divination date is for the Lesser Cold and Great Cold days, which are solar dates toward the end of the lunar calendar. Great Cold almost always falls in the twelfth lunar month, usually January 11/12 or 21 by our calendar, whereas Lesser Cold, which is December 26/27 or January 5/6 in our system, falls as often in the eleventh lunar month. When either falls in the twelfth month, wind or snow occurring on that day will signal losses to domestic fowl and livestock. The compiler finishes with the comment that these prognostications are based on experience. This information may not be part of the "orthodox" knowledge contained in the guides to monthly duties that the Ming government issued every year to its subjects, yet, he insists, the reader will find that "it is often remarkably accurate in predicting flood or drought and disaster or good fortune," which is essential "as an aid to agricultural affairs."[53]

The household encyclopedias that flooded the market in the Wanli era also offer methods for determining when disasters will strike. The prolific Fujian publisher Yu Xiangdou gives his readers a method for calculating the occurrence of natural disasters in his 1599 encyclopedia, *The*

Correct Source for a Myriad Practical Uses (Wanyong zhengzong). The method works with the calendar. Every day in the Chinese calendar is named using two characters, the first from a series of ten characters known as the Ten Stems, the second from a set of twelve known as the Twelve Branches. These are combined to produce a repeating cycle of sixty days.[54] The divination method involves figuring out which of the Ten Stems falls on the day known as Lichun, Starting Spring. This is another of the solar dates in the Chinese lunar calendar, which in our calendrical terms falls either on January 26/27 or February 5/6. If the Ten Stem character for Starting Spring is *bing* or *ding,* there will be a massive drought that year; if *ren* or *gui,* then the rivers will flood. If Starting Spring falls on a *wu* or a *ji* day, you will lose your fields. *Geng* and *xin* days herald peace and prosperity; *jia* or *yi* days promise a bumper harvest.[55]

If we were to apply these predictions to the first twenty years after *The Correct Source* was published, starting with 1600 and going to 1619, the findings are not persuasive.[56] The Starting Spring method indicates that 1600 should be a year of great drought, yet the rain in Fujian that year was so intense that city walls crumbled and bridges collapsed. Yu Xiangdou does get lucky for 1603, which his method predicts to be a flood year, for a tidal wave inundates the southern end of the Fujian coast and drowns over ten thousand people.[57] Other than that, all the predictions fail.

I have done this test not to mock the superstitious tendencies of the people of the Ming. My point is simply to show how exposed they felt to the unpredictability of the natural world. Dogged by fluctuations in the weather and haunted by the specter of famine, people of the Yuan and Ming turned to any device to help them prepare for the worst and give them hope that not all was beyond their control even if, at times, it seemed that it was.

The Good Years

Despite the severe bouts of bad weather and colder temperatures, the four centuries of the Yuan and Ming were far from being an uninterrupted run of unmitigated disasters. Bad weather brought flood and famine, the dire effects of which fed an undercurrent of anxiety that the future was unmasterable, but there were also the good years. Population grew, especially through the Ming, and the state more than covered its operating costs until the border wars of the final quarter-century drained the treasury. Private wealth accumulated, often to astonishing levels, and even modest prosperity touched the lives of the majority. The years were not always fat, but nor were they always thin.

To find the good years, we can simply reverse the chronology of catastrophes just recited and see what we find. The contrast is instructive. The Yuan dynasty did begin in the midst of a global downturn in temperatures, yet we see Khubilai right up to his death in 1294 ruling through a run of mostly good years. The good years paused during the Yuanzhen Slough (1295–1297), which began the year he died, but it was the least severe of the nine sloughs. So environmental conditions were reasonably good right through the first half of the Yuan, giving the Mongols half a century of relative prosperity. Later commentators thought so. "The Yuan dynasty from the moment that Khubilai unified the realm experienced peace for sixty or seventy years," wrote Ye Ziqi a century later in his commonplace book, *The Scribbler.* "Those who lived were properly nourished, and those who died were properly buried." It was an age that "truly deserved to be called prosperous."[58] Two centuries after Ye Ziqi, Jiao Hong passed much the same judgment on the early Yuan in his commonplace book, *Comments from Jade Hall:* "Of the nine generations of Mongol rulers of the Yuan, Khubilai was the wisest, as his rulership of his era fully attests."[59] All this changed with the coming of the Taiding Slough (1324–1330). The 1330s, and again the 1350s after the Zhizheng Slough (1342–1345), could be counted as relatively good spells, though famine and flood did become persistent problems. The conditions for agricultural bounty gradually eroded, undercutting the prosperity of the early Yuan and opening the way for the popular discontent and armed rebellion that brought down the dynasty.

The opening phase of the Ming dynasty was blessed with good years that lasted far longer than the sunny first half of the Yuan. There was the fierce epidemic of 1411 and the terrible famine of 1434, but recovery from both was reasonably swift. Not until the onset of the Jingtai Slough (1450–1455) in the dynasty's ninth decade did disaster strike the realm. It would not be a gross exaggeration to suggest that no major dynasty before the Ming or after was so blessed in its early phase.

Conditions remained poor for a decade beyond the Jingtai Slough. In the 1470s they improved, and the years were reasonably good up to the 1490s, when the Hongzhi emperor found himself having to deal with dragons and other disturbances. But conditions did not become really dire until the Zhengde Slough (1516–1519). That slough slowed the economy, but it did not derail it. The Ming then entered its warm phase, which with some interruptions—most strikingly the Jiajing Slough of 1544–1546—stretched on into the 1580s. If we bracket that catastrophe, the reigns of the Jiajing and Longqing emperors (1522–1572) experienced a run of remarkably good years. These two reigns, as we shall see, were the years when the Ming economy grew to such a degree that regional economies around the globe swung into its orbit.

The good years ended with the debacle of the first Wanli Slough (1586–1588), a crisis that touched almost every corner of the realm. Recovery followed, but conditions remained unsettled: dipping at the end of the 1590s, recovering again, then sinking into the second Wanli Slough (1615–1617). Good weather returned through the 1620s, though court politics and border incursions, as we shall see, eroded whatever positive impact it might have had on people's lives. Good fortune began to fade at the end of the 1620s, disappearing utterly during the Chongzhen Slough (1637–1643).

The nine sloughs were dramatic waves of crisis for both the Yuan and Ming, but they were more like punctuation to a four-century text in which good years outnumbered bad than the full story. With the exception of the second century, which escaped sustained calamities, the people of these dynasties experienced roughly seven good years for every bad year. Over a stretch of a hundred years, thirteen were evil and the rest were good. The path that the people of the Yuan and Ming traveled was thus crossed by many shadows, but each time they reemerged into the light—except for the final years of the Ming, when all was in shadow.

4

KHAN AND EMPEROR

THE YUAN and Ming dynasties span a significant shift in the politics of emperorship. Centuries of custom regarded the emperor as Heaven's son. By Chinese kinship rules, he was the only person entitled to approach Heaven in sacrifice and to communicate with it regarding the needs and hopes of Humans. He reigned as Heaven's deputee, but as Heaven remained aloof from the Human world, he could expect little help from that quarter. Instead, he ruled with the aid of an elaborate structure of personnel: some appointed on the strength of their literacy (as bureaucrats recruited through the examination system), some by virtue of their military skills (as officers and soldiers), and some because of castration (as eunuchs, which allowed them to work inside the imperial household without compromising the sacred imperial patriline from father to son). Each group represented interests that were rarely congruent with the interests of the other groups; indeed, within each group there could form factions locked in mortal competition. The power of these factions was such that it was a rare emperor who could assert his right to rule against them. Additionally, he was constrained by centuries of ritual institutions that laid down the rules regarding who could be emperor (the eldest son of the previous emperor) and how he could conduct himself. Even the most forceful emperor worked within what we might think of as a constitutional arrangement: a set of rules difficult to change or evade.

A Mongol khan was not a Chinese emperor. He occupied a very different position within his polity and enjoyed a very different relationship with those he ruled. He commanded soldiers and used a variety of means to recruit military supporters and allies, but he did not have a cadre of bureaucrats on whom to rely, nor did he entrust his household affairs to eunuchs. To whom he was born and the order in which it happened mattered, but not as much as it did for an emperor. Primogeniture might determine succession, but not without other factors coming into play. A khan (or in the case of Khubilai, a *khaghan* or khan of khans) had to triumph over his rivals and then have his leadership ratified by election at the assembly of nobles, the *khuriltai.*[1] On his death, it was expected that his sons, and often his brothers' sons as well, would fight among themselves for the succession. The practice of brothers competing with one another to succeed their father is known as tanistry, and was as much a legitimate principle of succession as was primogeniture among Chinese emperors. When it involved fratricide, and that was often enough, it is called bloody tanistry.[2]

Primogeniture aims for stability, the ideal at the heart of the imperial system and the condition that sedentary agrarian society favors. Tanistry does not aim to replicate the polity; it seeks to reinvigorate it. In a nomadic economy, reinvigoration was the key to surviving on the thin ecology of the steppe grasslands. A khanal succession was often predictable, but it was seldom orderly. Imperial succession sought to reproduce the same arrangement of power in every generation, whereas tanistry sought to reorganize it, and in so doing could come at the cost of political dissolution.

Khubilai's grandfather, Temüjin, better known by his title Chinggis (Genghis) Khan, understood the power of tanistry to tear apart the empire he had built. For that reason, he summoned his sons on his deathbed in 1227 and told them a fable. There were two snakes, he said, one with a thousand heads and a single tail, and one with a thousand tails and a single head. A cart approached, and the first was crushed because the heads all wanted to flee in different directions, whereas the other was able to slither away from the cart's wheel without difficulty.[3] His sons took the suggestion and agreed that the eldest, Ögödei, should succeed Chinggis as great khan. There would be no contest. This succession was the exception, for Ögödei was succeeded in 1251 not by his son but by his nephew Möngke after a round of fighting among the cousins. Möngke was succeeded not by his son but by his brother Khubilai, who through politics, force, and luck managed to outflank his two surviving brothers and win election as great khan of all the Mongols in 1260.

A Chinese emperor was not required to beat out his brothers, nor prepared to do so. Such a test was not only impossible but, from his perspective, only proved the barbarity of those who handled successions in this way. It placed them firmly beyond the pale of Chinese civilization. Ritual precedence, not force of arms, must prevail—or that was the theory. Practice did not always follow this rule, especially after the Yuan dynasty. That was because the Ming emperors could no longer be just Chinese emperors in the Tang or Song manner; they were also inheritors of the Mongol khanal tradition. Later political philosophers such as the Ming loyalist scholar Huang Zongxi (1610–1695) liked to isolate the Yuan dynasty as the most significant rupture since the unification of China by the Qin dynasty in 221 BC. After these two great upheavals, declared Huang, "nothing at all survived of the sympathetic, benevolent, and constructive government of the early sage kings."[4] This is hardly an objective account of Chinese imperial rule, but such was not Huang's purpose. It was, rather, to tar the Manchus, his unwanted Inner Asian overlords, by damning the previous round of Inner Asian conquerors.

Despite the politics of his opinion, Huang was not wrong in regarding the Mongol invasion as a major rupture, as modern scholars have argued.[5] But it was not a rupture that, once completed, somehow closed and disappeared when the Ming took over. The Mongols changed the course of imperial history precisely because those who defeated them perpetuated some of their norms. Conduct (such as tanistry) that had once been regarded as incompatible with Chinese traditions became, if never quite openly, a Chinese norm. The charismatic Mongol khan having become a routine Chinese emperor, the potential nonetheless lingered for the Chinese emperor to claim the charisma of a Mongol khan, and to act without regard for the constitutional constraints of emperorship. A few Ming emperors, most notably the first (Hongwu) and third (Yongle), would do just that. The rest found the predicament of being a supreme ruler who did not enjoy untrammeled power too great to do anything but sink into the political morass known as the court.

The Yuan State

Tanistry animated the outwardly chaotic succession of Yuan emperors. Khubilai Khan won the Mongol title of great khan in 1260, declared the founding of a Chinese dynasty, the Yuan, in 1271, took the Chinese imperial title of Zhiyuan, and was given the equally Chinese posthumous title of Generational Patriarch (Shizu) when he died in 1294. His longevity meant that the succession jumped a generation. He was succeeded by his grandson Temür (1265–1307), who was not the eldest but won the *khuriltai* election against his elder brothers Kammala (1263–1302) and Darmabala (1264–1292). When Temür died, the succession slid sideways to Darmabala's line in the person of his son Khaishan (1281–1311). On Khaishan's death it went to his brother Ayurbarwada (1285–1320), and thence to Ayurbarwada's son Shidebala (1303–1323). Shidebala was murdered in 1323 and the succession went back up a generation and over to the line of Temür's eldest brother, Kammala, who was Shidebala's uncle. It stayed there for five years with Kammala's son, Yesün Temür (1293–1328). When Yesün Temür's young son, Aragibag, was put on the throne in the fall of 1328, he held it for only two months before Darmabala's line took it away from him. For the next five years, two generations of Darmabala's heirs fought among themselves for supremacy. Toghön Temür finally emerged in 1333 to become the last, and longest reigning, emperor of the Yuan dynasty (see appendix). Counting from the final year of Temür's reign, 1307, to the first year of Toghön Temür's, 1333, ten khans sat on the Yuan throne in seventeen years—and it would have been eleven were it not that Tugh Temür became emperor twice.

Beneath the swirl of succession lay the edifice of the Yuan state, which Khubilai and his closest advisors, Chinese for the most part, shaped around Chinese institutions. Even before the Yuan, his father, Ögödei, had begun moving the Mongol state away from the practice of deriving its revenue through a combination of trade and tribute established by his father Chinggis. He saw the greater benefits of direct rule and direct taxation, and Khubilai consolidated the trend. His descent from the steppe to incorporate China was not a sudden move but a development following on the earlier phase of state formation that stood behind him. Khubilai was also pushed to build something like a Chinese imperial state by the costs of administration and an avaricious Mongol aristocracy, whose support depended on the rewards they received from him. He needed to conquer the Song in order to stay in business.

Khubilai's first move in this direction was to relinquish the old Mongol capital in Karakorum for a site further south. In 1256 he deputed his monk-advisor Zicong to plan the construction of a new capital city, known in Chinese as Shangdu—and romanticized in English as Xanadu. Once he had eliminated his rivals, he gave Zicong a second commission nine years later to build him a new capital 300 kilometers further south on the site of what had been the Southern Capital of the Liao and Jin dynasties, Beijing.[6] Henceforth, with the exception of the first fifty years of the Ming, Beijing became the permanent capital of China. Khubilai engaged a Muslim architect, Yeheitie'er, to design his new capital on a breathtakingly grand scale, which he did by combining Mongol elements of military display with traditional Chinese architectural forms. The result was a Mongol-Chinese hybrid that turned its back on Song architectural style and yet has come to be regarded as typically "Chinese." This move confirmed Khubilai's decision to be emperor as well as khan. Every summer he returned to Shangdu, his summer capital, to escape the heat of the North China Plain and go hunting. The hunt provisioned the court with food, provided military exercise for his troops, and gave Khubilai the opportunity to display his skills as a Mongol horseman and hunter. Liu Guandao's painting of Khubilai hunting in 1280 captures the emperor engaging just this display (Fig. 6).

Moving to Beijing entailed setting up a state that, like the architecture of the city, wove Mongol elements into a Chinese design. The administration of the realm was handled by the Central Secretariat (Zhongshu sheng). Its offices were situated just outside the south gate of Khubilai's residence, the Imperial City. The secretariat advised the emperor on policy matters, drafted laws, and supervised the traditional set of six ministries: Personnel, Revenue, Rites, War, Justice, and Works. The Ministry of Personnel made appointments, did personnel evaluations, and recommended promotions and demotions. Revenue was charged with conducting censuses and collecting taxes. Rites handled the court's heavy round of ceremonial obligations as well as the supervision of the examinations (when these were resumed) and the conduct of foreign relations. The Ministry of War, a civil rather than a military agency, oversaw the organization, supply, and training of the army as well as the operation of the courier system. Justice handled the administration of law, and Works managed state construction and civil engineering projects from walls to canals to imperial tombs. The central government also included a Privy Council to oversee military affairs. Khubilai was confident of being able to keep a close eye on matters in the capital, but to ensure that provincial administrators did not compromise Mongol interests, he appointed Mongol envoys, *darughachi,* to the Branch Secretariats.

Khubilai was concerned lest power slip from Mongol into Chinese hands, which was one of the reasons that he preferred to recruit his officials by recommendation, which established a bond of personal obligation, rather than through anything so dangerously anonymous as the tra-

Fig. 6 *Khubilai on a Hunt* by court painter Liu Guandao, 1280. Khubilai Khan was sixty-four when he posed for this painting and was at least as huge as he appears here. National Palace Museum, Taiwan, Republic of China.

ditional Chinese examination system, which selected officials on the basis of merit alone. Nonetheless, he did make efforts to project an appearance of moderation and benevolence to assuage his conquered subjects. According to an official report, though there is no way to verify this, he ordered only ninety-one executions between the years 1263 and 1269. Even by today's standards this shows remarkable penal restraint.[7] Such gestures were remembered. Ye Ziqi, the author of the early Ming commonplace book, *The Scribbler,* praised Khubilai's reign as a time of "lenient punishments and light taxes, when soldiers were withdrawn and rarely used."[8]

Among the Chinese institutions one might have expected Khubilai to do away with, but which he didn't, is the Censorate. Censors were appointed to monitor the conduct of officials and emperors alike. Their purpose was to ensure that the constitution not be undermined or compromised. It was at times a dangerous job, at other times a powerless one. The Yuan emperors declined to invest the Censorate with any serious authority. The agency was there merely to make sure officials did what the emperors told them to do. Shidebala, the fifth emperor, was the exception. In February 1323, as part of his program to recruit more Chinese into his administration to counterbalance the power of his Mongol opponents, he promulgated a set of guidelines urging censors to root out corruption within the government. The coup that assassinated him seven months later was led by none other than the Mongol censor-in-chief, who distrusted his cozying up to the Chinese.[9]

Nicola di Cosmo, a historian of Inner Asia, has argued that Khubilai's success in incorporating China produced an edifice that was "fundamentally flawed. It was ethnically disharmonious due to institutionalized racial divisions. Moreover, the metropolitan government was plagued by the proliferation of an extraordinarily wasteful central administration made up mainly of service agencies for the emperor and his entourage" (this is where we would find Marco Polo, among so many others who entered Khubilai's administration from outside). Di Cosmo also notes that "the Mongols' attitude to governance remained erratic and negligent, and some of the characteristic features of the inner Asian political tradition—such as principles of inheritance, privileges granted according to race and lineage affiliation, and the partnership between sectors of the central government and merchant organizations—remained very much in evidence."[10] It was an edifice that could be sustained as an imperial dynasty over the long term only with the unwavering support of Chinese officials—whom the Mongols never fully trusted—and the stabilization of the rules by which authority was constituted and transferred—which was never attained. And so it fell, but after standing for a full century.

Ming Autocracy

Zhu Yuanzhang came to power as the Hongwu emperor by overthrowing Mongol rule, though most of his efforts before 1368 were devoted to battling contenders up and down the Yangzi valley. He called his dynasty Ming ("bright"), a fire-word that cosmologically should succeed Yuan, a water-word meaning "primal origin." (Matter—like human affairs—was understood to cycle through the five phases of metal, wood, earth, water, and fire.) The word also signals his debt to the Manichaean cosmology of the struggle of the forces of light against the forces of dark that was part of the religious ideology of the rebel group with which Zhu associated early in his rise to power.

Zhu proclaimed that his mission was to rid China of its Mongol influences and restore Song models. It was a tale that comforted his Confucian advisors and may have played well to popular ethnic chauvinism, yet his new regime tended to reproduce the Yuan practices with which he was familiar.[11] The effect was to blend the traditions of the Mongol khans and the Song emperors into a new model of rule that the eminent twentieth-century historian of this period, Frederick Mote, half a century ago labeled "despotism." Mote believed the Song to be the point of origin of Chinese despotism, but he singled out what he called Mongol "brutalization" for "destroying much of the restraint" built into the imperial constitution, thereby opening the way for Ming despotism. "The brutalized world of the Yuan is of significance chiefly because it was the world in which the first generation of the Ming's rulers and subjects alike grew to adulthood, and in that way it helped to establish the tone and character of the Ming dynasty."[12]

Mote put forward this hypothesis about the Song-Yuan origins of Chinese despotism to challenge the Cold War sinology of Karl Wittfogel that prevailed in the 1950s and 1960s. Wittfogel argued that Asia was locked in an eternal condition of despotism extending from the deepest past to the present. The concept was something European intellectuals devised in the seventeenth century to characterize regimes in western and south Asia. Not until the eighteenth century did the term grow to include China, and eventually to denigrate China as despotism's highest stage.[13] This designation was an element in Europe's bid to construct an ideology of imperial hegemony that justified its own imperialism, the after-effects of which continue to shape our ideas and expectations about China.

In the next generation of Ming historians, Edward Farmer shifted the discussion of Ming government from the language of "despotism" to that of "autocracy," defined as "the enhanced concentration of power in the imperial institution."[14] The definition isolated autocracy as a system of political organization embodied in institutions. The issue was not the Mongol habits of Khubilai Khan, nor the ferocious personality of Zhu Yuanzhang, nor the authoritarian nature of Chinese society, but the institutions they introduced to preserve power. The term may be useful as a quick snapshot of a shift in the power of the emperor away from the constraints imposed by the procedures of his court and the expectations of his people. But there have been few rulers in history who actually ruled alone (the "auto" of autocrat). Even for the most determined of Ming emperors there were "practical limits" to his power, as Mote observed.[15]

One of the institutions we might consider as a brake on autocracy is law. The Ming founder ordered a code of law promulgated in the first year of his dynasty. Its statutes were intended to regulate the conduct of officials and commoners, however, not to constrain the emperor. Zhu Yuanzhang did not regard himself as bound by his own code. He found it frustrating that the Code did not provide the severity of sentence he thought appropriate for particularly heinous crimes, so he created what he called "law beyond the law." In the mid-1380s he gathered these extrajudicial judgments into a series of published texts called the *Grand Pronouncements (Dagao)*. The *Grand Pronouncements* became effectively a second law code. Zhu ordered his officials to respect the spirit of its decisions, but he carefully insisted that no one but he could use its sentencing guidelines. A judge who wished to invoke a penalty from the *Grand Pronouncements* was obliged to reduce it by one degree.[16]

Zhu was not unaware that people regarded the new penalties as excessive, but he believed them necessary in the degraded times in which he had come to power. When his minister of justice asked for permission to upgrade some of the punishment in the Code to the level of those in the *Grand Pronouncements* a year before Zhu's death, the emperor turned him down.[17] It was a moment when he might have allowed the sentencing prerogatives of a khan to overwhelm the legal institutions of an emperor, but he let the moment pass. Still, the regulations and institutions Zhu Yuanzhang put in place conspicuously lacked the sense of reciprocity between ruler and advisor or between ruler and people that Confucianism praised as an essential principle of good government. His idea of government hollowed out the Confucian moral tradition and left it with only punishments to maintain the health of the administration. Zhu's death the following year ended one of the most peculiar eras in Chinese history, when autocracy, even despotism, came closest to being realized (Fig. 7). Even though he left behind the command that "not a single word" of his instructions could be altered, his descendants could not sustain the autocratic constitution he sought to impose.[18] They had

Fig. 7 Portrait of Zhu Yuanzhang (r. 1368–1398). Was this unflattering sixteenth-century ink portrait of the Ming founder meant to caricature the man or to capture his unusual character? National Palace Museum, Taiwan, Republic of China.

to tamper with the spirit of his laws, even if their letter was sacrosanct. Tampering after all is what goes on in all political systems when what happens veers away from what should happen: when an emperor refuses to perform his duties, for example, or gets captured in war, or fails to produce a son. Such crises could be got around only through some degree of fudging and analogical reasoning. Yet because the rules were so inflexible, every crisis turned into a succession crisis, and every resolution came at a cost to the system's capacity to respond to future threats. Rather than seek to understand Ming rulership by tracking its political operations in normal times, we will see it in the context of five of the dynasty's major crises. The first occurred after only a dozen years of the founder's reign.

The Hu Weiyong Purge

The first major constitutional crisis shook the new dynasty to its foundations in its thirteenth year. Zhu Yuanzhang charged his prime minister, Hu Weiyong, with plotting to assassinate him, conspiring with hostile foreign forces (certainly the Japanese, perhaps the Vietnamese, possibly even the Mongols), and wanting to replace Zhu's dynasty with his own. Whether the charge was true or concocted is impossible to say, as all the records concerning Hu Weiyong have been doctored or destroyed; but it is possible. The last straw was Hu's failure to report a tribute embassy from Champa (Vietnam). Receiving tribute was an imperial prerogative, not the right of a prime minister. The charge, woven from thin shreds of errors and suspicions, may have had a basis in fact, but it played to the paranoia of someone newly on the throne. The *History of the Ming Dynasty* is uncharacteristically sparse on the details, presumably because the details were thoroughly suppressed at the time and unavailable to Qing historians.[19]

Underneath the charge, however, lay the constitutional reality of a prime minister's power. As the head of the civil bureaucracy, he had the authority to make appointments, and therefore could place his supporters in all the important posts, effectively building an entire administration that was staffed independently of the emperor's choices. As that was his job, Hu was not necessarily exceeding his mandate on any point except the reception of an embassy, assuming this charge was even true. What Zhu could not tolerate was the possibility that his prime minister was excluding him from administering the realm hands-on. He eliminated Hu, and in time anyone ever connected to him. Zhu himself estimated the number of victims to be 15,000 people. A string of purges followed over the next fourteen years, leading to the further execution of some 40,000 state officials at all levels. The purge of the 1380s was the most horrendous bloodbath of civilian violence in human history to that time, and inflicted a far greater trauma on the educated elite than anything the Mongols had ever done.

The cancellation of the position of prime minister revised the constitution by weakening it. The operation of the government now depended entirely on the intelligence and capacity of whoever happened to be emperor. As the Central Secretariat, which the prime minister had headed, was dismantled and permanently banned, there was no coordinating agency above the six ministries to channel civil affairs. The Chief Military Commission and the Censorate were similarly dismantled as coordinating executive agencies and split into smaller units without overall direction. As Charles Hucker summarized the outcome, "After 1380 Ming government was structured so that no single appointee could possibly gain overall control of the military, the general administration, or the surveillance establishment. Executive control remained in the hands of the emperor."[20]

This is the harshest, and perhaps most realistic, assessment we can make of the purge. It certainly crippled the Ming government for some years and put officials on heightened vigilance for two decades, yet it might be possible to see it from another perspective. Rather than simply a vengeful attack by a paranoiac who suspects that every action he does not himself initiate is a threat to his personal power, might we consider the purge as a decisive break with the practices of crony administration and favoritism that gave the Yuan government such a bad name? Given the loss of sources, it may be impossible to take this proposal any further. But if we think the continuities from the Yuan to the Ming were truly significant, then we might have to consider a purge on this scale as an adjustment of this legacy rather than simply being a case of an emperor trying to be more khan than a khan.

The concentration of administration in the single person of the emperor proved to be beyond even someone as driven as Zhu Yuanzhang. He was soon obliged to reintroduce coordinating agencies, though he did so in an ad hoc fashion. The most important step came two years after the purge in 1382, when he selected some low-ranking officials in the Hanlin Academy, the office in which imperial edicts were drafted, and appointed them as "senior scholars" to advise him—individually though, not as a collective. These grand secretaries, as we now translate their title, gradually formed a kind of cabinet. The Grand Secretariat was not firmly written into the organizational structure so as to replace the old Central Secretariat, but by the 1420s it had gained authority as the executive agency of the regime. Thenceforth the senior grand secretary was in charge of state administration, though only in close consultation with the emperor: a de facto prime minister, but without the authority or independence that Hu Weiyong had enjoyed.

The "Pacification of the South" Crisis

The death of the energetic founder in 1398 did not immediately plunge the Ming into its first succession crisis, but it came soon enough. Zhu Yuanzhang had appointed his eldest son as crown prince and successor, but that son predeceased him in 1392, so by the rules of primogeniture he designated the crown prince's eldest surviving son, Zhu Yunwen, as the heir apparent. Zhu Yunwen's enthronement in 1398 as the Jianwen emperor was a disappointment to his numerous uncles, many of whom had hoped to succeed their father. They rightly feared that the officials around Zhu Yunwen would draw him closer to the Confucian model of ruler and away from the martial traditions of his family.

The chief contender among the uncles was Zhu Yuanzhang's fourth son, Zhu Di (1360–1424). Zhu Yuanzhang had moved his capital south to Nanjing, the economic heartland of his new realm, in 1368, but he needed a strong defensive force in the north. He regarded his fourth son as a man who could hold back the Mongols, and so enfeoffed him on the site of the old Mongol capital, then Dadu, now Beijing. Barely a year after the Jianwen enthronement, Zhu Di asserted independent military control of the north and then opened a three-year civil war with his nephew. Officials loyal to the Jianwen emperor rallied to the defense of the reign, especially in Shandong, which formed a buffer between the two powers, but they could not prevail against Zhu Di's ruthless military might.[21] Zhu Di's forces descended on Nanjing, and the capital fell almost without a fight. The imperial palace was torched, and Jianwen was believed to have died in the fire. Rumors of his escape nonetheless continued to circulate for decades.

Zhu Di justified what he called his "pacification of the south" by playing the north-south card. The south, he claimed, had fallen into evil ways, and he was rescuing the dynasty from the misguidance of self-serving officials. Jianwen's (convenient) death was an unfortunate turn of events, not a regicide, he claimed. Four days after the fire, he ascended the throne as the Yongle emperor—and not as his nephew's successor but as his father's. He ordered that Jianwen's reign be wiped from the records. The enthronement in 1402 was officially dated not to the fourth year of the Jianwen reign but to the thirty-fifth year of the Hongwu reign, even though his father had died in the thirty-first year. This way of calculating meant that there had been no coup, and that son was succeeding father. Not until 1595 would the four Jianwen years get put back into the official chronology of the dynasty.

Zhu Di made the mistake of thinking he could win over Jianwen's chief advisor, Fang Xiaoru (1357–1402). A staunch conservative who believed that the only way to improve the world was to restore ancient ways, not to adapt to contemporary practices, Fang could never agree to the replacement of the legitimately enthroned emperor he had advised, let alone countenance an imperial succession from a nephew to an uncle. According to the principles laid down by Yongle's father, the only possible successor to Jianwen was Jianwen's son. Yongle decided to test whether Fang Xiaoru would submit to his rule by ordering him to pen the edict authorizing his succession. Fang would not. He threw his brush to the ground, declaring that he would prefer to die. Yongle acquiesced and ordered him executed by *lingchi* or "death by a thousand cuts."[22] Yongle later grandly declared that "I use only the Five Classics to rule the realm," yet the doctrine of moral reciprocity between superiors and inferiors that animates Confucian ethics is hardly in evidence in his reign.[23] When the emperor's authority was absolute, the virtue of informed loyalty dwindled to the vice of abject subservience. Fang Xiaoru was but one of many court officials who would pay dearly for choosing loyalty to the dynasty over subservience to the man who happened to hold power at any one moment in time.

Fang was not the only victim of what Yongle termed his "pacification of the south." The coup was followed by the execution of tens of thousands in a bloodbath that rivaled the worst of his father's purges. A second founder in the mold of the first was on the throne. The autocratic turn in Chinese politics has been laid at the feet of the Mongol emperors who ruled Yuan China, yet emperors Hongwu and Yongle were decisive in hollowing out the core Confucian values of obligation and reciprocity that the Ming regime might have nurtured in the restoration of the old imperial system.

Yongle completed his revamping of the regime by moving the central administration back north to the old Mongol capital, Beijing (Northern Capital). Serious construction began in 1416, and on October 28, 1420, the city was formally designated as the dynasty's capital. Nanjing (Southern Capital) was demoted to the status of secondary capital.

The stench of illegitimacy being strong, Yongle had to mobilize every device he could think of to mask it. One was to move the capital from Nanjing to Beijing. This located the political center in his base area; it also implicitly aligned the Ming with the warrior traditions of the Jurchen Jin and the Mongol Yuan rather than with the literati traditions of the Song. Yongle looked to Khubilai for his models. Another was to announce to the maritime world, as Khubilai had done, that he was now emperor. This he did by dispatching a series of trusted military eunuchs at the head of diplomatic missions to tributary states around Southeast Asia. Best known of these is the Muslim eunuch who led six of these missions, Zheng He (1371–1433). Zheng's first expedition between 1405 and 1407 got as far as the southwest coast of India before turning back to the Ming. Five more expeditions followed in 1407–1409, 1409–1411, 1413–1415, 1417–1419, and 1421–1422, all on a grand scale and at great cost to the Ming state. With Beijing simultaneously under construction, the financial burden was severe. A seventh was ordered, but after a fire burned three buildings in his newly constructed palace in 1421 (conventionally a sign of Heaven's disapproval), Yongle suspended that plan and died before another could be launched. Under the advice of fiscally responsible officials, subsequent emperors agreed that the state should stop building the enormously expensive "star-guided rafts" or "treasure ships," as his great vessels were variously known, and put the state's resources to better use than sending out inflated overseas missions to impose the dynasty's will and acquire mere exotica.[24]

There has grown up a curious urge to view Zheng He as China's antecedent to Christopher Columbus: as an intrepid explorer who, were it not for the penny-pinching bureaucrats back home, would have gone on to discover the Americas long before Columbus. This urge has led to much fantasizing among amateur historians, but it is based on a fundamental misunderstanding of the voyages of both Zheng and Columbus. Columbus was not an explorer. His voyages were vehicles of speculative commercial investment aimed at establishing direct trade links with China, a notion he was able to float to his backers in part on the basis of his reading of Marco Polo. He sailed west because he thought this route would get him there. His principal backers were the king and queen of

Spain, who were able to raise the funds by siphoning off some of the money expropriated from Spanish Jews in the great expulsion of 1492. Their interest in the voyages was principally financial, not diplomatic or political or intellectual. Columbus was crossing the ocean to trade, not to colonize, though he did leave groups of men behind to establish toeholds to supply future voyages.

When Columbus is viewed this way (rather than as the heroic explorer who "discovered" the Americas and changed the world), Zheng He begins to emerge from the mist of misrecognition as more his opposite than his avatar. Zheng's purpose was diplomatic: a mission to declare to all tributary states known to China that Yongle was now the emperor and that they should send him tribute to acknowledge the fact. He took with him a sizeable military force to make sure that the rulers on whom he called did not refuse his command, but he was not intent on conquest. China had an interest in lubricating commercial links throughout maritime Asia, and its fleets helped Chinese merchants to enlarge their trade circuits, but the voyages were not targets of investment. Nor were they expected to produce the stunning returns in gold that Columbus promised, and consistently failed to deliver, to Ferdinand and Isabella. Finally, Zheng's ships did reach places to which no Chinese officials had ever traveled, notably on the east coast of Africa, but they were sailing known routes that Muslim traders in the Indian Ocean had long been using. Chinese mariners may have been unfamiliar with some of these places, but they were not in any sense "discovering" them. They were simply adding them to the roster of states that should acknowledge Ming suzerainty. Zheng He was not an explorer-entrepreneur out on the ocean to discover the world; he was an imperial servant sent to get the one thing that his usurper-emperor craved: diplomatic recognition. This was political theater, and no less important for being so.[25]

One rumor circulating around the expeditions was that Yongle believed the Jianwen emperor might have escaped the fire and fled abroad, and that Zheng He's mission was to find him. The rumor is improbable, though what gives it plausibility is the fact that the Yongle emperor was indeed on the outlook for reports of his nephew, of which over the years there were many. One of the last was in 1447, when a monk over ninety years of age told someone he met on the road between Yunnan and Guangxi that he was the dethroned emperor. When this boast reached the ears of local officials, the monk was arrested and dispatched to the capital. Under torture the poor man admitted that he was a commoner, that he had entered the clergy in 1384, and that he wasn't Jianwen. Shen

Defu, who records this story in his commonplace book, *Unofficial Gleanings of the Wanli Era,* notes that the interrogators should have figured out the old monk's age and done the math before they started in on him. Zhu Yunwen was born in 1377 and would have been seventy in 1447: the monk was twenty years too old to be Jianwen. "The imposture should have been immediately obvious," Shen caustically observes.[26] The monk died after four months in prison. A dozen other monks with whom he associated were defrocked and posted to the miserable life of border guards in the far north, the direction from which the next crisis would come.

The Tumu Crisis

Yongle passed his throne to his eldest son (the Hongxi emperor), but it went in quick succession to that son's eldest son (the Xuande emperor) and then to his eldest son (the Zhengtong emperor). This third successor, Yongle's great-grandson Zhu Qizhen, was enthroned in 1435 at the age of eight, bringing to the fore a distinct weakness in the imperial constitution: having to enthrone a child whenever early mortality and the rules of succession demanded it. The three grand secretaries whom the boy-emperor inherited from his father, all surnamed Yang, provided stability during the emperor's minority. When the Zhengtong emperor came of age at fifteen, however, the three Yangs lost control of the court to the Directorate of Ceremonial, the leading eunuch agency in the palace. The eunuchs were more willing to let Zhengtong follow his whims, the most calamitous of which was his desire to command a military expedition against a Mongol incursion across the Great Wall. This desire led the dynasty into a constitutional crisis that no one anticipated: what to do when an emperor is taken hostage by a foreign power?

The provocation was a three-pronged invasion of north China by Esen, who had recently reunited the Mongols under his command. Zhu Qizhen appointed the Prince of Cheng, his half-brother Zhu Qiyu, as regent while he was on campaign and left Beijing the following day to lead what has been called "the greatest military fiasco of Ming times."[27] Things went from bad to worse in a matter of weeks. The moment of reckoning came when Esen caught up to the imperial entourage at Tumu, a courier station between the inner and outer Great Walls, as the emperor was scuttling back to the capital. The Chinese force, refusing to negotiate, was massacred and its entire senior officer corps annihilated. On September 3, 1449, Zhengtong was taken hostage.

The court back in Beijing had two constitutional options at this point: accept the hostage-taking and negotiate the emperor's return, or revoke his emperorship and install a new emperor. As the latter course was being contemplated, two more options opened up: enthrone the captured emperor's one-year-old son, or allow the succession to go sideways to the half-brother who was already acting as regent. An infant emperor put the imperial system in its weakest position—not what the Ming needed to get out of the crisis. A compromise was worked out, and twenty days after Zhengtong's capture, Zhu Qiyu was enthroned as the Jingtai emperor, and Zhengtong's baby boy was named heir apparent. The year 1449 was still recognized as the fourteenth year of Zhengtong, but 1450 would be the first year of Jingtai.

This move could be interpreted as a coup d'état; Zhengtong certainly thought so. Even the street urchins of Beijing caught the fragility of the situation, for they were soon chanting this nursery rhyme:

> Raindrop, raindrop,
> City God, Earth God.
> If the rain comes back again,
> Thank Earth God for bringing rain.

The words sound innocent enough. After all, the realm had for the last dozen years been in the grip of a dry spell, and rain was on everyone's mind. But it was all a tower of puns. Read "rain" as "give" in the first line and as "emperor" in the third (all three are homophones), change "drop" to "brother," switch the left-hand radical on the character for "city" to turn it into Jingtai's title, Prince of Cheng, and the poem turns into a satirical comment on the succession:

> Gave his brother, gave his brother,
> Land to the Prince of Cheng.
> If the emperor comes back again,
> He'll have to hand it back again.[28]

The enthronement of Jingtai had canceled out the hostage's value. A year later Esen released his useless prize in exchange for a weak Chinese promise to reopen border trade. The Jingtai emperor would not permit Zhu Qizhen to enter Beijing until he had clearly renounced his claim to the throne, which he duly did. To secure the throne for a new line of succession stemming from himself rather than his brother, Jingtai deposed his nephew in 1452 and installed his own son as heir apparent. The boy died within a year, during the Jingtai Slough, and his death was interpreted as evidence of Heaven's displeasure. The loss of his son intensified the pressure on Jingtai to reinstate the nephew he had deposed a year earlier.

The reign did not go well. This is hardly surprising, for the entire Jingtai era was consumed by the Jingtai Slough. The weather was abnormally cold throughout the era, going from killing drought through the first three years to hopeless waterlogging for the last two. One official in 1454 openly attributed the evil times to Jingtai's disrespect for the correct order of succession. "Restore the Prince's status as heir apparent; secure the great foundation of the realm. If this is done, then gentle weather will fill the realm and the disasters will end of their own accord," he told the emperor. Jingtai was furious and ordered the man taken out and executed, but next day a sandstorm blew up and shrouded the capital. Fearing that this was Heaven's rebuke, he stayed his hand.[29]

Jingtai fell ill in the winter of 1456–1457 and was too sick to attend the morning audience on New Year's Day. A coalition of civil and military officials took matters into their own hands, releasing Zhu Qizhen from his house arrest and putting him on the throne, to the complete astonishment of those who showed up at court for morning audience. Rather than resume his reign as Zhengtong, which would mean having to erase Jingtai from the history books as his great-grandfather had erased Jianwen, Zhengtong took a new regnal title, Tianshun, Going Along with Heaven's Will. By March 14, the deposed half-brother was dead, though whether by illness or murder, no one can say.

Historians have regarded the installation of the deposed Zhengtong emperor as Tianshun in various ways, as "the coup d'état par excellence of Ming history," a "grave violation of ritual propriety," and "an act of political opportunism that unleashed a flood of profiteering and office seeking."[30] But if we think of it as a round of tanistry in the best Mongol tradition, then however offensive it may have been to Chinese ritual propriety, it was not inconceivable that an ambitious prince should seize power from his half-brother when that brother was weakened. To remove all evidence of impropriety, Zhu Qizhen over the next four years purged the conspirators who had put him back on the throne. There is, in fact, nothing in the eight years of the Tianshun era that the Chinese imperial tradition looks back to with pride.

The Great Ritual Controversy

The next constitutional crisis was also a succession problem that led to what some regarded as another grave violation of ritual propriety. After Tianshun, the eldest surviving son followed his father through the reigns of the Chenghua, Hongzhi, and Zhengde emperors, until Zhengde died in 1521 without an heir. The question of who would succeed him was compounded by the problem of his having been unsuited to the job in the first place. Enthroned at thirteen and married off at fourteen, Zhengde showed little interest in either his empress or his empire, and left his first half-decade of rule to his senior eunuch, Liu Jin. Liu's rapacious administration outraged the civil bureaucracy, against whom the eunuch operated a reign of terror.[31] The situation reached crisis proportions in 1510 when the Prince of Anhua, Zhengde's great-great-uncle, rebelled. The rebellion was suppressed, but the politics it unleashed generated a rumor that Liu was plotting to assassinate the emperor. It was enough to turn Zhengde against his eunuch, for whom he ordered a three-day execution by slicing, though Liu expired on the second day. The next decade went almost as badly, leading to a second rebellion, this time by the Prince of Ning, Zhengde's great-uncle, in 1519. The military campaign against the prince led by the philosopher-statesman Wang Yangming (personal name Wang Shouren, 1472–1529) ensured the continuation of the hapless emperor's regime.

Two years later, Zhengde fell ill, allegedly in consequence of falling drunk out of a boat while fishing the previous autumn (dragons at work?). He had fathered no child, despite having collected a harem in the hundreds, though that may have only been an elaborate hostage scheme to squeeze money out of their families.[32] Leaving no heir and no written instructions for his succession, officials in his inner circle started backing rival candidates among his uncles and cousins. Yang Tinghe (1459–1529), the most powerful official at court, prevailed in putting forward a thirteen-year-old cousin, Zhu Houcong. The rules did not permit succession by a cousin, so the Ministry of Rites recommended that a posthumous adoption be arranged with Zhengde's long-deceased father. This would make the boy Zhengde's younger brother, and younger brothers were permitted to succeed older brothers.[33] As Zhu Houcong's deceased father had been the younger brother of the previous emperor, this was acceptable within Chinese adoption practices, which allowed a brother's son to be adopted as his uncle's legal heir.

Zhu Houcong was cleared to become the next emperor, but then he threw a wrench into the works. He wanted to establish his legitimacy differently, and in a way that no one at court had considered. Rather than make himself his imperial uncle's lineal heir, he wanted to elevate his deceased father posthumously as emperor (as well as elevate his mother to the status of empress dowager). That way he could descend from his father ritually as well as biologically. Here was a constitutional conundrum of the first order. Instead of pretending that the new emperor descended directly from the ruling branch of the imperial family, it would move the succession to a collateral branch. But this opened the possibility of a constitutional challenge by other members of the Zhu family, and in the wake of two princely rebellions, no one wanted that to happen. The dominant faction at court tried to wear the young emperor down, but Jiajing remained adamant. The issue became a crisis over the question of which rituals the young emperor should perform for his biological father: the rituals due to a natural father, or the rituals due to an uncle? This question divided the bureaucracy for close to a decade in what came to be called the Great Ritual Controversy.

Matters came to a head on August 14, 1524, three years into Jiajing's reign. Several hundred officials staged a demonstration outside a gate of the Forbidden City. They would not permit the emperor to treat his ancestry as a private matter; it rested at the very heart of the entire edifice of Ming imperial succession. One hundred and thirty-four demonstrators were eventually arrested, eight of whom were sentenced to life exile and the remainder punished with lesser sentences, including flogging. Sixteen of the flogged men died as a result of their beatings. The protests continued and a second round-up led to three life exiles and another death by flogging. But still nothing was resolved. The following spring Hou Tingxun went further than his colleagues: he put his criticism into print. Arrested and tortured, he was later released on the plea of his twelve-year-old son, and eventually rehabilitated and reappointed—only to be cashiered and reduced to commoner status later, on corruption charges.[34]

The emperor's opponents did not occupy all the high ground. He had his supporters, and not just among the predictable crowd of ambitious outsiders trying to get ahead. That support came from none other than the followers of the great mid-Ming philosopher Wang Yangming. Wang had distinguished himself by suppressing the Prince of Ning in 1519, though that victory gave him such credit that jealous rivals did everything to block his entry into the inner circle at court, leaving him sidelined for the rest of the Zhengde reign. The Jiajing emperor appointed him Minister of War, but the death of Wang's father in his first year in office obliged him to go on mourning leave. He remained sidelined, and silent, during the 1524 protest. In June 1527, Jiajing reinstated him as Minister of War and ordered him to lead a campaign against a rebellion in Guangxi province near the border with Vietnam. The campaign was immediately successful—his mere arrival terrified the rebels into surrendering without a fight—but Wang fell ill and died on his way home before he could have a direct impact on the Jiajing court. His final campaign may have helped to push his disciples' faction into leadership at court the following year.

Wang avoided expressing a direct opinion on the matter of Jiajing's ritual obligations, but he was not unsympathetic to what he saw as the emperor's natural impulse of filial piety, which he believed was the true foundation of moral action.[35] Those who opposed Jiajing tended to accept the authority of Song Neo-Confucianism and its reverence for textual precedent, while those who supported him believed that right moral action depended on ethical intuition. Jiajing's desire to honor his father was not a self-contained constitutional issue, therefore, but became the first clear declaration from the highest authority in the land that the individual had scope for a degree of moral independence from precedent. Wang's philosophy of innate ethical intuition was no longer just an academic project but had found political footing. It was not Wang's theoretical position so much as the ascendance of his own supporters at court that enabled Jiajing to win the day. Even so, the rise of Yangming Neo-Confucianism was intimately tied to the constitutional politics of that succession. As a result, as James Geiss has pointed out, "Wang's teachings became known throughout the empire in a very short time and remained a subject of great interest and contention into the seventeenth century."[36]

The "Foundation of the State" Crisis

The legitimacy of dynastic succession depended on naming the correct heir apparent. Failing to do so was an opening for a constitutional crisis, which is precisely what happened during the reign of Jiajing's grandson, the Wanli emperor (r. 1573–1620). Wanli was the eldest surviving son of the Longqing emperor, who was Jiajing's eldest surviving son by one of his concubines. Longqing reigned for only six years (1567–1572) before succumbing to a premature death, which left the throne vacant for the child who became the Wanli emperor. Once he reached majority, the new ruler had to think about designating his own heir apparent. The problem that drove his ministers and himself to distraction was that he did not want his own eldest son to succeed him. He preferred his third eldest, Zhu Changxun, the son of his favorite concubine, Lady Zheng. This preference undammed an endless stream of trouble.

The struggle began in 1586 when he conferred the august title of Imperial Consort on Lady Zheng. He also wished to elevate her son to the status of heir apparent. The court divided, as it had done over the elevation of Jiajing's father. As the heir ensured the legitimacy and therefore the continuity of the reigning dynasty, he was spoken of as the "foundation of the state," and so this controversy became known as "the foundation of the state controversy." Unlike the battle over Jiajing's choice of father, however, there was no moral high ground on which Wanli's supporters could stand. This was purely a matter of whether to go along with or defy the incumbent's personal preference, whether to insist on preserving the correct ritual order or permit the rules of succession to change. The actual roadblock had little to do with either prince, and everything to do with Wanli's anxiety to please his concubine.

Officials at court were well aware that Lady Zheng was the emperor's favorite, and both she, and they, took advantage of her connection to Wanli. In 1588, Lü Kun (1536–1618), a prominent official engaged in a variety of social renewal programs, published a small book of stories celebrating the virtuous conduct of women through history. *Models for the Inner Chamber (Guifan)* reached Lady Zheng's attention, and she commissioned a new enlarged and illustrated edition that included additional stories about twelve more models of virtuous conduct, the last of which was none other than Lady Zheng. The new edition featured prefaces by her uncle and brother broadcasting her patronage. A clear harem bid for power, it unleashed a firestorm of protest aimed at Lady Zheng but naming Lü Kun as the target.

After three more years, Wanli finally caved in to his advisors and agreed that his eldest son be installed as the crown prince. Lady Zheng kept up the pressure on behalf of her son nonetheless, provoking a violent round of denunciations and arrests two years later when a pamphlet appeared on the streets of Beijing accusing her of recruiting nine top officials to launch a coup against the "foundation of the state."[37] Still, Wanli kept up his campaign to cast Lady Zheng in the best possible light.

The campaign reached something of a high point in May 1594, when Wanli leveraged her donation to a famine relief project in Henan to force all capital bureaucrats of the fifth rank and higher to donate their salaries alongside her gift.[38] Regardless of who did what at this point, the ritually correct choice held. The irony of the whole business is that within nine days of his enthronement as the Taichang emperor in 1620, the eldest son fell ill and died within a month, possibly poisoned by wrongly prescribed drugs to treat an illness. The throne did not slide sideways to Zhu Changxun but passed unremarkably to Taichang's own remarkably incompetent eldest son, who became the Tianqi emperor. In the densely bureaucratized world that the Ming had become, a tanistry challenge did not arise.

Wanli's struggle to install his third son as the crown prince lacked the violence of earlier succession struggles, but it had the significant impact of embittering the emperor, who thereafter more or less absented himself permanently from court.[39] The entire body of capital officials showed up for daily audience every morning, as they were required to do, but the throne was usually vacant. The business of the realm could only proceed off-stage, as it were, by enfranchising eunuchs, deferring to grand secretaries, and otherwise improvising procedures in a way that would have baffled and outraged the dynastic founder. Wanli was less a dragon master than a harem master. The fiscal historian Ray Huang, who is largely responsible for our image of Wanli as an isolated and frustrated emperor imprisoned by his bureaucracy, has blamed the stalemate of the Wanli era on "the impossible conditions imposed on the monarch, conditions which had grown haphazardly from circumstances rather than from design. Although an autocrat by definition, the emperor had no legislative powers. Although the final arbiter, he had to operate in a legal haze."[40] Wanli's contribution to this stew was inaction—which constrained whatever his officials tried to do, but also limited his own possibilities.

Thus alienated from the seat of imperial power, officials did what members of badly managed institutions usually do in order to create pathways for action: they formed factions. The faction that emerged early in the seventeenth century to fill the leadership vacuum took its name from the Donglin or Eastern Forest Academy, a private scholarly institution in Wuxi that a group of Jiangnan intellectuals revived in 1604 as a forum for public discussion. A magnet for young men pursuing common causes, the Donglin party soon emerged as the chief counterweight to the eunuchs of the Imperial Household, and effectively to the emperor himself. Had Wanli been able to develop the political skills and moral authority of some of his early ancestors, he might have been able to break the stalemates in his own court and get on with ruling. But how was someone whose entire life was defined by the four walls of the Forbidden City to learn such skills or locate sources of authority other than the record of his birth?

The Predicament of Loyalty

It would be ludicrous to cast the emperor as the tragic victim of his own autocracy. If there were any tragic figures, they were the men like Fang Xiaoru who identified moral issues over which they could not compromise, and were willing to court his displeasure and their own destruction in their defense. It might be better to think of the politics of the Ming court as a matter of bargains, not tragic flaws. Most everyone understood the loyalty clause in autocracy's deal between ruler and minister: only the minister can be at fault. Whether the ruler acts well or badly is of no account, because he is essential to the system, the foundation of the nation, and the only sure guarantee of the survival of his dynasty. Loyalty produced a situation that was a predicament both for the ruler who wished to override the real constraints on his power but could not figure out how to do so, and for the official who believed that there were constitutional principles higher than the obligation to serve the emperor but who could not hold to the one without suffering for breaching the other.

An elderly Buddhist monk by the name of Huilian, whom Lu Rong, author of the commonplace book *Miscellany from Bean Garden,* interviewed in the mid-fifteenth century, understood perfectly what the imperial predicament meant for officials. Huilian had been called to Nanjing early in the century to work on the *Great Encyclopedia of Yongle (Yongle dadian),* a massive scholarly compilation of all known texts. The project ran from 1405 to 1408 and consumed the energies of some of the best scholars in the realm. Huilian subsequently retired to Lu's home county, and was over eighty when Lu met him.

"In the Hongwu era," the monk told his student visitor, "licentiates who served in office suffered so much, lived in so much fear, and engaged in so much mental effort for the court. Yet in the end, if you were guilty of a slight misdemeanor, the light punishment was being exiled to a border garrison, and the heavy was having your corpse desecrated. Those who came to a good end amounted to only two or three out of ten." Such was the fate of those who served the Hongwu emperor. Huilian then offered his lesson, one that was not what we might expect—nor Lu Rong, for that matter.

"Many were the gentlemen throughout the realm on whom the dynasty turned its back," Huilian conceded, "but no gentleman of that time turned his back on the dynasty." It was of no account whether the emperor mistreated his officials. What mattered was their willingness to submit to whatever the emperor dished out. Far from being a tragedy, their submission proved their unquestioning loyalty. The young people of today, Huilian declared, fall short of the autocratic ideal. Even though "the emperor is magnanimous and the law loosely applied" these days, young men refuse to serve, acting on no principle other than to save their own skins.[41]

There was no room in Huilian's world view for autonomy. This did not mean that the idea, or the ideal, was absent. But it was an ideal that the individual who pursued it, whether by serving with detachment or by withdrawing completely from public office, had to do quietly. And it is quietly that we can detect it lurking behind much of the verbiage of the era. Take, for instance, the official essay that the unstoppably prolific writer from a Huizhou merchant family, Wang Daokun, was commissioned to pen to preface the published honor roll of graduates of the Guangdong provincial examinations in 1582. The commission demanded an expression of impeccable loyalty, and at first glance this is all it appears to be. Wang casts back over the history of the dynasty to single out several emperors for special praise. Hongwu "received Heaven's mandate and revived the realm"—no debate there. Jiajing "reilluminated and perfected the realm" when he "emerged from the capital of Huguang," his princely fief. "His greatness superseded the achievements of his predecessors, his brilliance lighting the land within the seas"—a bit of a rhetorical stretch for a ruler who tied his court in constitutional knots for years. Wang does allow that "two or three illustrious gentlemen" were obliged to "keep their own counsel and withdraw to their home areas" but casts no further shadow on the terrible relationship the emperor had with many of his officials.

As for the currently reigning emperor, Wang writes that when Wanli issued "ten thousand policies from his great height, the power of his wisdom spread to the four quarters. Many scholars have shot to sudden prominence as though they had taken to sea or flown into the air, like clouds hanging in Heaven, surpassing the phoenix," as the Guangdong graduates undoubtedly hoped to do also in due course. Wang characterizes their graduation year, 1582, as a year in which talent flowered as never before "by the grace of the Son of Heaven. Today our native places throng with scholars who are responding to the benefits of this glorious age." The pivot of this wondrous vision is Wanli. "What regulates the times of Heaven and aligns the skeins of Earth is rooted in the virtue of the emperor and finds its response in human achievement," Wang intones. "That so many scholars have benefited from this has not been seen in almost a thousand years."[42]

It is difficult to imagine a more craven account of the Jiajing and Wanli reigns. But everyone knew how matters really stood, and we would be mistaken if we allowed ourselves to be fooled by the lofty language. Something else is going on here. The heroes in this drama—the phoenixes rising into the air—are not the emperors, in fact, but the scholars. Against great odds, they have devoted themselves to decades of study and given themselves to the service of not just the ruler but the Heaven and Earth he is supposed to anchor. The job of the emperor is merely to rest at the base of the system, embodying virtue and letting those with real competence get on with the tasks of ruling. The ritual ruler could not be trusted to rule well. Or so Wang Daokun and the brightest of his generation, despite their platitudes, actually believed.

The Wanli emperor was on his guard against such ambition down among the provincial graduates.[43] But he was less alert to the possibility that his own ambition for autonomy from the ritual prescriptions of the dynasty would undercut his own authority, as he would learn in trying to elevate the constitutionally incorrect son to the post of crown prince. It was his predicament that the constitution of the dynasty had to prevail over his personal preference—just as it was the predicament of his officials to submit to the requirements of loyal service rather than pursue the higher moral purposes for which their education trained them.

5

ECONOMY AND ECOLOGY

THE prosperity Marco Polo found in Khubilai's realm amazed him. The population was "enormous," he declared; the countryside "pleasant," the towns "fine," the fields "well-tilled," the quantities of merchandise "enormous." As his barge floated down the Grand Canal, he laid eyes on "so many towns, villages and scattered homes that one might say that the entire route is inhabited. Nowhere along the whole journey is there any lack of provisions, such as rice, wheat, meat, fish, fruit, vegetables and wine, and the like, all of which are bought very cheaply." In town after town, the people "live by trade and industry. They derive great profit from their thriving commerce. They have ships in plenty."[1] This was productivity beyond European imagining.

Two centuries later, in 1488, the Korean official Ch'oe Pu was washed ashore in a storm on the Zhejiang coast. His overland journey home took him up the same Grand Canal, and there he experienced some of Polo's amazement. "Learned men and gentry abound," he wrote while being ferried up the Canal where it crossed the Yangzi delta. "All the treasures of the land and sea, such as thin silks, gauzes, gold, silver, jewels, crafts, arts, and rich and great merchants are there." Around the city of Suzhou, "market quarters are scattered like stars" and "the people live luxuriously. There are solid rows of towers and stands." At the wharves along the canal, "merchantmen and junks gather like clouds."[2] For these splendid sights he can think of no real-world counterpart, only the fanciful descriptions of ancient palaces in Tang poetry.

Marco Polo has been accused of exaggerating his reports of the Yuan, on the one hand; Ch'oe Pu on the other is regarded as a sober and careful reporter of Ming life. Yet each tells his reader roughly the same story: that this realm was a place of great prosperity, a land of good order, an economy of surplus. Natural disasters took their intermittent toll, but did so without reducing the economy down to subsistence. People produced a surplus that sustained polities and societies at a material level well above what people in Korea or Venice, or anywhere else in the world, were experiencing during these centuries. That would change as Europe plunged into its early-modern transformation at the end of the Ming, but the full consequences of that transformation lay far in the future.

The Mixed Economy of an Agrarian Empire

The majority of subjects living in the agrarian empire of the Yuan and Ming worked as grain farmers growing millet, sorghum, and wheat in the north, rice and winter wheat in the south. Polo thought the "well-tilled fields" worth remarking on, and had the impression that "no land is left idle that might be cultivated." What Ch'oe noticed, though, were the "solid rows of towers and stands" and the "merchantmen and junks" tied at the wharves. So did Polo, for that matter, whose eyes kept wandering to the "enormous quantities of merchandise" and the "ships in plenty." All these economic activities across the spectrum from plowing to manufacturing to trading were essential for supporting this empire. Their interdependence made it possible for commerce to expand, agriculture to become prosperous, and cities to grow. Commercialization was not a one-way arrow, but it flew often enough to make people, especially in the late Ming, think it was.

The state collected the bulk of its revenues from the production of grain. The Yuan state in 1299 reported collecting 12 million *shi* (1.15 billion liters) of grain in taxes.[3] Assuming that an adult male consumed six *shi* (570 liters) of grain per year, then the registered population of sixty million in 1330 would have needed to produce 360 million *shi* (34 billion liters) of grain annually for subsistence.[4] Not every registered person consumed at the level of an adult male, but then the real population may have been larger by half as much again. If we permit these distortions to cancel each other out, then the amount levied suggests a collection rate of 3.4 percent—which turns out to be in line with the rate at which Chinese states have usually taxed grain yields.[5]

The levy for 1393 tells a very different story. The Ming state reported collecting 24,729,450 *shi* (2.65 billion liters) of rice and 4,712,900 *shi*

(447 million liters) of wheat.[6] Combined, the grain levy that year amounted to 3.1 billion liters, two and a half times greater than the levy in 1299. If the demand for grain was roughly the same as it was in the Yuan, then the Ming state was collecting grain tax at a rate of 9.1 percent, noticeably higher than in the Yuan.[7] What accounts for the jump? The difference may indicate that the Ming state was more effective than the Yuan state in extracting revenues from the economy, and it probably was. It could also indicate that population was higher than recorded, yielding greater aggregate tax revenue. It might also suggest that the state was levying grain from a more productive economy, and this is likely as well.

The early Ming administrations made great efforts to stimulate grain production by relocating population to regions where land had fallen fallow during the inter-dynastic war. Zhu Yuanzhang's ideal was that every farming family should have 100 *mu* of land (6.5 hectares), which was considered necessary to support a large family in the north and abundant for a family working in the more intensive agricultural economy of the south.[8] By the sixteenth century, that ideal had been reduced to 50 *mu* in north China.[9] One frustrated northern student declared in the 1620s, when his father gave him that acreage, that this reduced figure was not enough. "How can a real man in this world possibly support himself on 50 *mu* of land?" He promptly sold the plot and joined the army.[10] In south China, many households scraped by on as little as 20 to 30 *mu*.

After the early Ming recovery, the productive economy consisted primarily of farmers growing grain to near maximum capacity, and the fiscal economy consisted of state levies collected almost entirely in grain. Once the economy was on its feet, however, the state began to shift its levies from kind into cash (a reform known as the Single Whip, to which we will return). As it did, the amount of grain reaching the center declined. One impetus for this change was transferring the capital to Beijing. By locating the capital in a northern agricultural environment that could not sustain its population, the government had on the one hand to intensify its efforts to amass grain there, hence the rebuilding of the Grand Canal. On the other hand, it recognized that feeding everyone in the capital and at the defense posts on the northern border was beyond its capacity to collect and distribute, and that it could handle that task more effectively by converting taxes to cash and using the cash to stimulate private commerce to meet these needs. This arrangement also left the grain itself in the provinces, which was available for redistribution from areas of surplus to areas of need. As a result, the provinces most distant from Beijing—Guangxi, Yunnan, and Guizhou—retained their tax grain for local needs rather than face the enormous cost of shipping it all the way to the capital for redistribution. Provinces less distant, but still at considerable remove—Guangdong and Fujian—were permitted to retain locally roughly two thirds of the grain they collected. In the vast grain-basket provinces of Sichuan and Huguang, regional officials were entitled to hold back about 60 percent of the tax grain.[11]

From the beginning, then, the Ming created a command economy but operated it in coordination with the private economy. This was more than a passive arrangement, for the Ming state provided conditions that facilitated production and trade far exceeding its own fiscal and monopoly interests. What distinguished the late-imperial Chinese state, in contrast, say, to European states in the same period, was the conviction that the state was responsible for the welfare of its people. Rooted in the Confucian principle of reciprocity, this conviction disposed emperors to show concern for their subjects' well-being, and officials to exert themselves in protecting and promoting it. As an imperial edict reminded officials in Beijing after a flood-induced famine in 1518, "Let every responsible official show extra care and concern, so that no one falls through the cracks."[12] This was not a mere gesture of benevolence, but a moral command and a bid for survival. Many an emperor neglected to meet this expectation, and many an official treated his own appointment as nothing but an opportunity to enrich himself, yet both failures prove the rule that a state that failed to nourish the people lost the mandate of Heaven.[13]

Much of the economic growth in the Yuan and Ming, though more so in the Ming, was private in organization and capital but public in terms of its operation within an infrastructure created and funded by the state. The state provided the transportation system by which commodities circulated. It eventually mandated the payment of taxes in silver, which then set the terms within which values were calibrated and exchanged. It operated monopolies on salt and precious metals and collected taxes on such a scale that its fiscal operations set directions for the private economy and shaped the livelihood decisions of ordinary people. It stored grain, and in times of famine it intervened in grain markets to ease the threat of dearth. It operated textile manufacturing workshops, mostly in the commercial centers of Jiangnan, to supply the needs of the imperial household.[14] It commissioned specialized workshops, such as the porcelain kilns at Jiangdezhen in Jiangxi province, or the tile kilns in Linqing, down the Grand Canal, to produce the objects needed to build and furnish the court. Finally, it provided the administrative and legal institutions needed to moderate conflict and manage economic disputes. Parasitic in appearance, the state helped constitute the economy in practice.

Transportation

Bulk goods travel more cheaply by water than by land, hence the importance of rivers and canals for transporting grain and other bulk commodities. Because the natural flow of water in China was from its western mountains to its eastern plains, the principal challenge for a state concerned to facilitate the circulation of commodities was how to arrange water transportation running north and south. The Grand Canal, the origins of which go back to the seventh century, would become the core of the Yuan and Ming state's north-south transportation strategy.

After Khubilai established his capital at what is now Beijing, he made do with shipping supplies north by sea. But the losses incurred from navigating the rocky coast of the Shandong peninsula, plus the vulnerability of slow barges to pirate attack, encouraged the court to consider other options for provisioning itself. One was to cut a canal across the base of the Shandong peninsula; this was tried and abandoned in the 1280s. The next was to revive the Grand Canal and extend it north from the Yellow River, where it had stopped in Song times, to Beijing. Construction costs were enormous, and maintenance was expensive. Consequently, the Yuan was not able to keep the Grand Canal in operation throughout the dynasty. Whenever silting, flooding, or warfare blocked it, Yuan officials shifted back to the sea route.

The challenge at the northern extension of the Grand Canal was to harmonize the flows of water in the Yellow River and the Grand Canal, which crossed each other. The Yellow River was prone to spilling its banks and changing its course, and every time it did so it threw the operation of the canal into havoc. Rechanneling the river was an expensive solution, requiring enormous numbers of laborers. The labor corvée of 1351 is often cited as the spark for the popular uprisings that eventually brought the Yuan dynasty down. Han Shantong, leader of a secret society known as the Red Turbans, was able to recruit a strong following among the 150,000 men dragooned into service that winter to redig the canal. Han was captured and executed, but his son, Han Lin'er, who later called himself the Lesser Prince of Brightness (Ming), escaped and became the figurehead of a rebellion that in time included Zhu Yuanzhang. The son died in 1366, leaving the way open for Zhu to take leadership of the rebellion. His choice of dynastic name acknowledges his debt to the Red Turban origins of the rebellion.

The Grand Canal lost its priority after Zhu decided that his capital would be in Nanjing. By 1391, the silted canal was unusable.[15] Yongle's decision to move the capital back to its Mongol site in Beijing forced the state once again to invest hugely in the Grand Canal.[16] The Ming proved better at sustaining their investment than had the Yuan, funding significant engineering improvements along the more difficult stretch over Shandong. The canal was reopened in 1415 and, with a few interruptions when the Yellow River shifted its course, remained in operation to the end of the dynasty. A huge contribution to the infrastructure that served to integrate the realm and its economy, the Grand Canal was also a huge burden. Maintaining it increased the already enormous pressure on the hydrological structures that sought to keep the water and the land each in its place. With every flood and every silting-up, the task only became harder.[17]

The size of the canal demanded labor and equipment on a scale equal to its requirements. By the mid-fifteenth century, 11,775 government grain barges were being hauled up and down the canal by 121,500 soldiers to keep the imperial storehouses in Beijing full. Actual collections fell short of official quotas, though what the soldiers who operated the barges carried north on their own account and sold into the private market made up for the shortfall. The imperial household also operated its own barges to supply the palace. These were said to number 161, of which fifteen were iceboats to transport fresh fish and fruit from the south. Accompanying these barges were some of the 600 skiffs called "fast-as-horse boats" (*makuai chuan*) that the Ministry of War operated to protect the imperial haul.[18] The official boats were outnumbered, however, by the private barges maneuvering their way along the crowded canal in the tens of thousands. As one mid-Ming writer pictures the scene, "With well over ten thousand barges shipping tribute grain annually from the south, the boat traffic north and south never rests for a day. As for the private boats and merchant barges, they are numerous beyond count, all following one another along this route."[19] The revival of the canal gave commercial transport the backbone it needed to connect the country rudimentarily into an integrated economy.[20]

Commercial travel could be a hardship nonetheless, and so commercial travelers relied on divination techniques to protect themselves against misfortune. "Master Yang's Inauspicious Days" provided a schedule for when no travel, packing, or trading should be transacted. It has been preserved in a merchants' almanac of 1635, *Warnings at a Glance for Merchants (Shanggu yilan xingmi)*.[21] The "inauspicious days" are numbered from the beginning of each lunar month and fall on every twenty-eighth day: the thirteenth day of the first month, the eleventh of the second, and so forth. Another popular list in the same almanac, "Monthly Auspicious Embarkation Days," relies on the sixty-day cycle of the Ten Stems and Twelve Branches. These lack the regularity of Master Yang's dates. For example, the sixth month could have ten days on which it is auspicious to start a journey, depending on where in the sixty-day cycle the month fell, whereas the third month could have only two. Every month also has five Catastrophe Days when a merchant would be a fool to launch any enterprise. In the first month they are the *si* and *hai* days, so in 1635, February 22 and 28 and March 6, 12, and 18 were embargoed. The sixty-day cycle also had its own good and bad days, regardless of the month in which they fell. The first *si* day of every sixty was mildly inauspicious to begin travel. The first *hai* day could be extremely auspicious, so long as you set out between the hours of 1 and 3 a.m., 7 and 9 a.m., or 7 and 9 p.m. On no account should you start your journey between the hours of 11 a.m. and 3 p.m.

Merchants engaged in these elaborate calculations to do what merchants have always had to do, minimize risk in an uninsured economy. This avoidance of risk penetrated all aspects of daily commercial life. Suzhou merchants, for example, would not use the words *fan* ("overturn") or *zhu* ("blocked"). As spoken Chinese has a high number of homonyms, this led to some intriguing word substitutions. The word for chopsticks is *zhu,* written with a different character but pronounced exactly the same as the *zhu* meaning "blocked." Since merchants most feared having their goods blocked, Suzhou people started using the opposite word, *kuai,* "fast"—the word we use today for chopsticks *(kuaizi)*.[22]

Cities

The growth of the economy stimulated the growth of cities as markets, as sites of manufacturing, and as the residential choice of elites. Beijing had the advantage of being a center of government, but it was also a commercial hub for the economy of the north, and had a population that must have exceeded half a million. The former southern capital, Nanjing, was on the same scale. One estimate suggests a population in 1400 of 700,000.[23] Further downstream on the Yangzi delta, however, lay the greatest cities of the age. Suzhou could boast of being the commercial and cultural hub of the empire, and must have had a population close to a million. The port of Shanghai, which also became a center for the vast cotton trade that emerged there in the fourteenth century, anchored a county of a million people, of whom at least a quarter may have resided in or around the city.[24] Hangzhou, though eclipsed compared to its heyday as the capital of the Southern Song, was still an elegant and wealthy city where anyone with money hoped to own a villa.

Commerce and administration flowed together in these major cities. In smaller cities, however, commerce came to outstrip administration as the engine of urban growth, though administrative duties were usually part of the city's origins. This happened to Linqing, situated at the point where the Grand Canal flows from Shandong province into North Zhili. Linqing had been a river county of no importance until the Yuan chose it as the new northern terminus of the Grand Canal. When that extension was completed in 1289, Linqing became the link between the southern economy and the northern state. The collapse of the canal later in the Yuan threw Linqing into obscurity until 1415, when the canal was put back into operation. Linqing's boost had already been prepared in 1369, when the county *yamen* was moved closer to Linqing lock, the difficult point where the canal connected to the Wei River that flowed northeast to Tianjin. It was chosen as one of five sites for granaries to hold the tribute grain coming north up the Grand Canal. It soon came to overshadow the other four, especially after 1450, when it was designated as the place where merchants who had contracted with the state to supply grain in exchange for licenses to sell salt (a system known as *kaizhong* or "border delivery") had to deliver their stocks and collect the licenses. Described as the strategic "throat" of all north-south commerce, Linqing was elevated in 1489 to the status of a subprefecture.[25]

Linqing's growth as a city, partly induced by the state and partly generated by private enterprise, continued in the sixteenth century. All northbound grain had to pass through Linqing, as did everyone coming on official or private business. The flow of traffic nourished the city. In the Chenghua era, local artisans previously obliged to travel to Beijing to perform their labor service were excused from that assignment and allowed to perform it in Linqing instead: the city's labor resources were effectively no longer held hostage by the capital. Beijing projects still needed to be supplied, but Linqing artisans could manufacture bricks and tiles at kilns in its suburbs rather than having to travel to the capital. Under these conditions, local industries handling the commodity traffic and supplying the needs of the region flourished. Shipwrights were busy in twenty-eight boat works in the northwest part of the city building canal barges, and textile merchants were operating 105 shops in the city by the end of the sixteenth century.[26] The scale of the market in cities like Linqing meant that manufacturing was no longer contained within household workshops but was enlarging into something closer to factory production (Fig. 8).

Despite the growth of private commerce, private industry in Linqing rested on foundations laid by state infrastructural investment. The same process shaped the growth of other cities along the Grand Canal.[27] But there are Ming cities where the factors were reversed: where private commerce led and the state followed. The Yangzi River, a natural artery, did not require the level of state investment that the Grand Canal did, so its riverports grew differently. Take for example Shashi, Sand Market, shortened from Shatoushi, Sand Spit Market. It is just that: a spit of river sand sticking out into the Yangzi River in Jialing county in the western part of Huguang. Jialing's county seat was almost 10 kilometers from the river, just far enough that it had to cede commercial priority to Shashi. The bulk trade on which Sand Market thrived was grain coming down from Sichuan. Sichuan merchants and bargemen were a large commercial presence in the city, but merchants came from all over the country. The city boasted ninety-nine merchant and artisan guilds by the end of the Ming.[28] The state could not ignore Shashi, but rather than elevate it to county status, which would have robbed Jialing of a major chunk of its revenue, it was given its own police station. The state was present in other ways as well, for central agencies handling taxes and materiel established offices here. The Ministry of Works, for instance, being charged with supplying logs for palace construction, set up a station in Shashi to monitor and tax the log barges floating down the Yangzi.[29]

As a commercial city, Shashi was more vulnerable to the booms and busts of trade than an administrative city. The poet Wang Qimao records this vulnerability when he returns to the city during the bad times of the Chongzhen Slough and recalls the prosperous world as he had known it in his youth:

Fig. 8 At the center of this industrial scene, four workers are operating a large grindstone. They may be grinding rice to make beer, or they may be grinding raw materials for the manufacture of porcelain. None of the other details conclusively confirms which is the case. Regardless, what the picture does show is the concentration of artisanal labor for industrial production. The original provenance of the painting, in the collection of the Harvard Art Museum, is unknown. Although dated to the Ming, the style suggests that it was painted in the Qing dynasty.

> I recall those years when I walked beside the jade-green Yangzi,
> When nothing there at Sand Ford was not in great profusion.
> Wine was poured in a thousand boats moored to sell their wares;
> Flowers bedecked the courtesans' brothels stretching on for miles.
> Now that the times are evil, few traders arrive from Sichuan;

> Now that the people are poor, few listen to the strum of lutes.
> Coming back to this marvelous place, I grieve to find it desolate;
> Sitting facing the empty woods, I glimpse a few geese in the sunset.[30]

The rise (and temporary decline) of Sand Market attest to the capacity of the commercial economy to determine how some cities emerged and developed largely without reference to the state. But cities of all types had to manage their affairs with little direction from the state. The state never flinched from asserting its monopoly over public affairs when it perceived a threat to its interests or revenue. That said, officials did have to devise ways of working with urban elites to arrange matters that fell outside normal procedures. The old models of village life were simply not viable when it came to handling the needs of hundreds of thousands of people, not just hundreds.

Consider, for example, the problem of urban fire prevention. Hangzhou was badly burned on May 4, 1341, though the loss of life (seventy-four) was modest compared to the destruction of buildings, tallied at 15,755 rooms. Whatever escaped the blaze of 1341 succumbed, however, to the fire on June 4, 1342. "From ancient times there had never been such a fire," the author of a 1366 commonplace book declared. "The accumulated splendor of several centuries was reduced to nothing in a single day."[31] Hangzhou was an administrative center, yet state mechanisms were insufficient to protect it from fire.

The prefectural capital of Yanping in Fujian further down the coast was also an administrative center, yet the state was a weaker presence, and the prefect had to coordinate with commercial elites to get anything done. Though not as densely built as Hangzhou, "throughout history," the compiler of its prefectural gazetteer notes, "fire disasters have burned large swathes of the city time and again." Yanping's vulnerability to fire was intensified by its being situated in "a hilly, cramped place where people live cheek by jowl." In 1575, the prefect ordered firewalls built. As this meant expropriating valuable urban property, he had to recruit five wealthy commoners to buy the land for him rather than force through an unpopular expropriation. Seven firewalls went up, one of which stood in front of the prefect's *yamen*. The walls were not continuous, however, and three years later over a hundred homes in the city center, including several *yamen* buildings, burned to the ground. The next prefect ordered the walls extended and connected, and even donated his salary to pay for it, but that was not enough. He had to go back to the original donors to organize a scheme to get the owners of the buildings where the walls ran to donate land for their construction. The original seven firewalls were expanded to nine. Lest future encroachment compromise the project, the prefect announced to urban residents that they had the right to appeal to his office. The plan worked, and the incidence of fires inside the city fell.[32]

The building of Yanping's firewalls appears at first glance to be the prefects' doing, yet urban elites played the decisive role in funding the plan, and probably in designing it in the first place. The gazetteer credits the prefects for troubling themselves over the welfare of their subjects, but the commoners who carried out the project—undoubtedly the leading members of Yanping's commercial elite—managed a task that a prefect on his own was ill-equipped to handle. The problems of cities were technically invisible to Ming administration, so urban elites had to solve them. We should be cautious, though, about concluding that urban elites took charge as they did in early-modern European cities. "The transition from an empire of villages to one of cities," writes urban historian Si-yen Fei, should not be seen only as "the triumph of commercial power that defied and eventually prevailed over the oppressive grip of the state." It came about through "the concurrent institutional reforms and cultural negotiations that bridged and reconciled the early Ming rural ideal and late Ming urbanization."[33] Urban people figured out how to make their cities work by adjusting formal administrative rules to the reality that their cities were no longer rural villages.

Taxing a Commercial Economy

Like most military occupations needing revenue, the Yuan regime had weak access to rural communities. To tax the countryside, it resorted to imposing fixed quotas and farming out these quotas to the highest bidder. A tax entrepreneur purchased the right to collect taxes from a specified area in return for delivering a set quota to the government. Whatever he collected over and above that quota, which could be considerable, was his to keep. The Yuan was also draconian in its labor levies, forcing farmers to abandon their agricultural work whenever their labor was needed regardless of the season. The Suzhou poet Zheng Yunduan (ca. 1327–1356) expresses this burden in the opening lines of her poem on the old story Chinese poets liked to tell of the woman who turned into stone waiting for her husband to return:

Her husband left for conscript labor,
Summoned to a distant corner of the realm.
He planned to return after three years' time,
Yet since his departure so many winters have passed.
She climbs a mountain that rears its precipitous peaks
And stretches her neck, watching for returning boats.
But no returning boat meets her gaze.[34]

The Ming rejected both the indiscriminate levy of corvée labor and the farming out of the land tax, which it regarded as not just immoral but inefficient. The one removed necessary field labor without regard to the rhythm of the agricultural year and without reasonable limits, while the other encouraged tax collectors to bleed the people, not to meet essential state expenses but for their own private gain. These models disrupted the base of society, interfered with the production of local wealth, and allowed public monies to flow into private hands. If either had any merit, and the Yuan clearly thought they did, it lay in lowering the costs of administering taxes. But there were other ways to do that. The model the Ming proposed was to turn over the administration of labor and land tax levies to the elders within the local community. This was the logic of the founder's registration system known as the *lijia*. Locals would know best who had labor available and who should pay how much land tax.

The vision of rural self-sufficiency free of landlord exploitation underpinning the *lijia* system—this was Zhu Yuanzhang's personal vision—was blind to the natural tendency of an economy receiving state investments to generate and concentrate wealth. In the fifteenth century this tension produced fiscal schizophrenia. The model of self-taxing autarkic communities drifted ever further from the reality of villages linked into commercial networks that encouraged production for the market.[35]

The tax system was slow to follow reality, as Gui Youguang (1507–1571) discovered when he took up his first post as a county magistrate in 1566 at the advanced age of fifty-nine (Gui had managed to fail the metropolitan examination with stunning regularity until 1565). "Although the tax system has set quotas, households still use the names of their ancestors of the Hongwu era," Gui wrote to his prefect, "so that when it comes time to collect taxes, everybody points to everyone else" for who should pay up. "On top of this, adjacent fields have been merged under a single owner, yet the records show the original households each owning a few bare *mu*," so everyone claims that their meager property puts them below the minimum for taxation. "And then there are the great households up in the hills who from one year to the next resist any sort of restraint," refusing to pay any taxes whatsoever.[36] Gui found the county's census records in a hopeless state. They showed that a heavily commercialized county had lost 20 percent of its population between 1488 and 1522 and thereafter not grown by a soul. Additionally, the figures suggested that women accounted for only 20 percent of the population.[37] Clearly the system was completely divorced from reality. The "real" economy—a money economy of commercial investment and financial concentration—had escaped entirely from the model of the agrarian economy installed back in 1368, and untaxed fortunes were being made.

The solution was to go with the changes and convert the two main resources of an agrarian economy, grain and labor, into monetary equivalents and work with these. Collect taxes in silver, and then use that silver to pay for the costs of administration. Magistrates understood that labor was more efficient when it was bought rather than dragooned. Better to hire a lockkeeper for four ounces of silver a year than summon and discharge a long list of corvéed laborers through the year who did not know the first thing about operating a lock and were as likely to disappear as do the job required of them. As this change made its way across the economy in the sixteenth century, particular levies in kind were funneled into a single levy in silver.

Moving the tax system from the founder's rural model of static communities to an economy of monetized exchange in the sixteenth century was the most important transformation of the Chinese economy prior to industrialization. We know it under the unusual term of the Single Whip. The term is a pun on the phrase *yitiao bianfa,* "the conversion of tax assessments into a single item." *Bian* means "convert" or "reform," but it also means "whip," so popular wit turned "one item" into "one whip," and the label stuck. This reform began piecemeal in the fifteenth century, to be formalized and extended under Chief Grand Secretary Zhang Juzheng in the 1570s—the same official who ordered a complete resurvey of all cultivated land in the empire in 1580. Zhang was regarded at the time as a monomaniac bent on increasing the reach of the state at any cost, particularly in matters involving state finances. Practically every commonplace book of the Wanli era includes some comment to this effect. "When Zhang Juzheng controlled the country, he rarely eased up on the implementation of the laws," commonplace writer Shen Defu declares. "Any theft of taxes in excess of four hundred taels entailed immediate execution."[38] Zhang is now seen as a visionary administrator who adapted the old agrarian tax system to a commercial model that laid the foundation for a modern economy.[39]

Chief among those adaptations was monetization. When the Ming founder imagined prosperity for his people, he imagined it in terms of closed communities of farmers, every household owning the land it tilled and growing the food and textile raw materials it needed to survive. Although that was never how the Yuan or Ming economy actually worked, the image captured an ideal of economic life in which money played almost no role. Where it existed, it was in the form of small bronze coins ("cash" or coppers) with a hole in the middle, a form of currency in use since the Warring States period. As simple necessities cost a few coppers, there was no need for a currency of higher value.

That was the theory; practice soon nudged it aside. As agricultural surpluses accumulated and more goods were turned into cash, larger transactions had to be paid for using coppers strung on strings of a thousand *(guan)*. These strings were cumbersome, and the coins themselves vulnerable in any case to being melted down for their metal or otherwise adulterated. This vulnerability put currency in chronic shortage, making the collecting of taxes an annual nightmare. Facing a copper shortage fully two decades before they founded the Yuan dynasty, the Mongols opted for the Song and Jin practice of issuing paper money denominated in "strings." The regime supported its paper currency with sufficiently large grain reserves that the paper kept its value until 1350, when—desperate for revenues to pay for its military costs—the government allowed the printing of money to outstrip the grain reserves that backed it up. By 1356, Yuan paper was so worthless that it ceased to circulate.[40]

The Ming followed Yuan precedent, issuing a currency known as Great Ming Precious Scrip, but it was inconvertible and inadequately backed up too, and so lost its value. Another form of currency was necessary to handle larger than everyday purchases, and silver moved in to fill the gap. The unit of silver was the tael *(liang)*, measuring 37 grams. Silver was not minted into coins, however, but stored in ingots that were weighed and stamped. (A judge sent down to Guangdong in 1601 was intrigued to discover that "the silver of the foreigners from the west is molded into the shape of coins and has fine writing on both sides.")[41] When a payment had to be made, the silver was weighed to the required value and the payment made literally in a lump sum. In this system of dual currencies, the exchange value of copper to silver fluctuated, depending on their relative availability, consumer confidence in their purity (copper coins were regularly debased), and government intervention in the currency market.[42] The debasement problem meant that several types of copper coins might circulate simultaneously at different exchange rates in the same market. The most valued copper coins were those minted in the Hongwu era.[43] They circulated throughout East Asia and were particularly valued in Japan. When an anti-Manchu resistance group on the Zhoushan Archipelago south of Shanghai in 1647 received a shipment of coins from supporters in Japan, they found they were all Hongwu coins. Typically, the Japanese melted down coins of all other reigns and minted them as their own, but not the Hongwu coins, which were hoarded for their value.[44]

As the economy grew, so did the values in which it traded, and hence so too did the need for silver. The government impelled this process forward in 1436 when it selectively converted some of its levies into payments in silver. The thirst for silver was made worse by the government's unwillingness to permit silver mining, for fear that the precious metal would pool in private hands and undermine the state economy. Only in the latter decades of the sixteenth century would that thirst begin to be quenched by massive silver imports from Japan and Peru. Without this silver, the dramatic incorporation of the Ming economy into a global trade (a later topic) might never have happened.

Food Supply in a Cash Economy

The principal product of an agrarian economy is grain. Those who dream of rural self-sufficiency, as Zhu Yuanzhang did when he imposed the *lijia* system, must also dream that every local economy down to the smallest community has the resources and conditions to plant and harvest grain in the amounts adequate to meet its own consumption needs. This is why grain, while the principal product, is not the only product of an agrarian economy. It is also why grain sometimes loses its original character as food for the farmer that grew it, and enters the sphere of commercial circulation as a commodity traded by other people.

Increasingly through the Yuan and Ming, this became the main story of the economy, especially as more people, first absolutely and then relatively, were living in cities or traveling the roads as merchants or staying in the countryside but producing other agricultural and handicraft commodities for the market. For these people, the food they ate was what they bought, not what they grew. It was a novel situation for political economists of the Ming, especially with the ghost of Zhu Yuanzhang breathing self-sufficiency down their necks. It was also a novel situation for granary administrators who no longer maintained grain stocks adequate to fend off sudden fluctuations in supply that, if not quickly corrected, could escalate to crisis proportions. This was the cost of commercializing an agrarian economy, dismaying artisans and economic conservatives alike.

To Ming officials, commercial grain had the baffling ability to move away from where it was in demand as well as to where it was in demand, depending on the aggregate power of that demand and the existence of competing prices elsewhere. When grain moved into deficit zones to feed the starving, the economy was judged to be doing what it was supposed to do. When the opposite happened, that is, when grain merchants removed grain from deficit areas in order to sell it in a region where the price was higher, commerce could be deplored. It was understood that laws of value were at work, as demonstrated by the inclination of humans to pursue their own interests. But officials did not regard working within those laws morally desirable when the outcome was local shortage.

Still, it was not always clear what to do in the face of dearth. A popular mantra of the time declared of the very worst famines that "there are no good policies for famine relief." Officials who quoted this mantra generally did so, however, to argue the opposite, which is that an effective official can always find ways to prevent a grain shortage from turning into mass starvation. "When dealing with famine, worry not that there are no brilliant policies," insists one veteran magistrate in his commonplace book *Random Notes on What I Have Seen and Heard;* "worry only that you have not concentrated your mind. If you concentrate your mind, then you will have brilliant policies."[45] This opinion echoes the subjectivism of late Ming thought, yet it resonates with an older conviction that no undertaking can succeed that does not originate in sincerity, and that true sincerity conquers all, even temporary glitches in local food supply.

The most basic of the "good policies" was to stockpile grain in government granaries, withdrawing it from the economy after bumper harvests and then making it available in times of dearth (an ideal expressed in Fig. 9). The Ming founder pursued this policy by requiring that four Preparedness Granaries be built in every county. Most magistrates built them, yet most of their successors understocked them, and most of these let them fall into disrepair, despite central directives to the contrary.[46] Decline was not just a matter of indifference, incompetence, or corruption, though any one of these three was enough to undermine a Preparedness Granary. Rather, it had to do with the nature of the political economy

Fig. 9 A magistrate distributing grain from a state granary in a time of famine. This illustration is taken from a late Ming novelization of the fourteenth-century opera *The Story of the Lute (Pipa ji)*.

within which grain is stored. Hongwu's assumption in creating the Preparedness Granaries was that they should be stocked with local harvest surpluses: the government was simply imposing savings in good years that the local farmers could then draw on in bad. However, as more and more people were buying their grain from dealers who acquired it from growers elsewhere, local food crises were quite as likely to be commercially induced as they were to be the effect of local harvest failure.

In a commercial economy, preparedness might take the form not of stockpiling grain but of stockpiling funds that could be distributed to the poor to pay famine prices, a move that should draw in commercial grain from elsewhere and cause prices to return to normal. The only problem with this "good policy" was that sometimes famine escalated to a scale beyond what grain merchants could manage. This seems to have been what happened during a province-wide famine in Shandong in 1307. The Yuan court's response was to distribute money, but it was soon discovered that there was no grain available for the starving to purchase, and no prospect of its imminent arrival. Fortunately in that case, an energetic official intervened and got the policy switched from relief in money to relief in government grain.[47]

Handouts that took the form of money were also dogged by the ease with which officials could pocket the funds. This happened during an epidemic that started in coastal Zhejiang the same year as the Shandong famine. By 1308 it had grown into a major disaster, and the stricken were starving. Regional Pacification Commissioner Tuohuancha duly applied to the court for relief. The court responded by ordering the local wealthy to contribute relief funds to Tuohuancha, which they dutifully did. As the collecting and auditing of these contributions were done locally, Tuohuancha saw an opportunity to make himself rich. His scheme involved dividing the funds into smaller sums and depositing these with many local officials, who were to hold the cash until he needed it. Tuohuancha allowed time to pass, then came back and collected it, weaving a paper trail of collections and disbursements too tangled to unravel.

In one county, he made the mistake of entrusting a sixth of the total to Assistant Magistrate Hu Changru. The son of an eminent Song official, Hu had been forced into service under Khubilai and just been demoted to this lowest of ranked posts for having opposed powerful interests in his previous post. Hu suspected that Tuohuancha intended to pocket the funds, so he immediately released the money he had been given to the local needy. Each recipient was required to fill out and sign a receipt. When Tuohuancha came back a month later to collect his money, Hu handed him the book of receipts.

"Your money is here," Hu calmly told him. Tuohuancha was furious. "You have the gall to do that? How dare you so recklessly disregard the order you received?" Hu was unfazed. "If for one day the people do not eat," he replied, "there will be deaths. The truth of the matter is that we have stopped short of the point at which deaths are being reported. The official documents are all there, so this can be verified." Tuohuancha had to swallow his anger and say nothing.[48]

Despite the ever-present possibility of corruption, most theorists of state management by the fifteenth century were prepared to argue that the granary system was an inadequate response to dearth.[49] The idea that the commercial economy does a better job of redistributing grain than does the state became a key element in the administrative reforms that Qiu Jun (1420–1495) laid before the Hongzhi emperor in 1487.[50] In the same vein, Lin Xiyuan (ca. 1480–ca. 1560), who undertook to reformulate famine policies in the sixteenth century, argued against the expectation that the state should provide relief. Rather, it should make use of existing commercial capacity in the private sector by engaging merchants to buy grain cheaply elsewhere and bring it to the famine region to sell. Grain merchants would be allowed to add a charge of two bronze coins to the selling price, half to cover transport costs and half as a commission. The state would provide the initial capital for the venture. Once the stocks were sold and the money returned, the total cost to the government would be nothing.[51] The state might intervene to strengthen demand with occasional cash inputs in order to get the process going but should otherwise rely on the market to meet subsistence crises.

What happened during the nine sloughs, however, exceeded the capacity of either the state or the market to respond. Often it truly seemed that there were "no good policies" for getting food to the people. As a commentator in Henan province lamented during the Jiajing Slough, "the lives of the people in former times relied on their ruler" who could be counted on to store grain for them, "whereas the lives of people of later times depend on Heaven alone."[52] When Heaven could manifest itself in the person of the emperor ordering relief for the stricken, everything still seemed right in the world. But when Heaven became the market, it was hard for people to imagine that anything stood between them and their eradication.

Through the last century of the dynasty, officials continued to experiment with new policies in the gray zone between the state and the economy, asking themselves how each could best be used and experimenting with the combination in practice. For the more enlightened among their ranks, famine relief was not simply intervening to address one particular crisis. It was part of a broad program of improving the lives of the people they called *jingshi* or "ordering the world." The phrase is the first half of the four-character phrase *jingshi jimin,* "ordering the age and aiding the people." Those who embraced this moral commitment, which we translate as "statecraft," understood that their role in serving the state was to mobilize whatever resources the state placed at their disposal to ensure that the people did not perish in hard times and flourished in good. Their commitment to public action was fundamental to the activist strain of Ming Confucianism that came to the fore in the middle of the dynasty. Its concern was the people, and its register of action was the economy. The power of this commitment was so strong that when a neologism was needed in the nineteenth century to translate the European concept of "economy," the phrase *jingshi jimin* yielded up the new word *jingji.*

The Perplexity of Prosperity

Statecraft activists argued for the need to intervene in the economy when survival was at stake. Significantly, their concern gained prominence not as the economy withered but as it grew. This development may indicate an awareness that commercialization can undercut survival just as readily as it can augment income. But it also points to a shift in expectations. Rural self-sufficiency was no longer the goal; a more commercially grounded prosperity was now what many people had come to expect. The state was generally on side with this goal, though as the economy grew, some of its officials felt concerned over what they saw as the unwanted effects of prosperity: social mobility, the decay of traditional customs, and the erosion of the established moral order.

Gu Qing was an official who felt himself at the frontline of such changes. He was a native of Songjiang, the prefecture covering the heavily commercialized eastern end of the Yangzi delta, which included the cotton industry around Shanghai. The death of Gu's father obliged him to return home to observe the obligatory twenty-seven months of mourning during the Zhengde era. While he was there, he got involved in compiling his prefecture's first gazetteer, published in 1512. Gu was clearly unhappy about local customs in Songjiang, for he opens that section of the gazetteer on the theme of *bian,* changes. "Observe the changes," he wrote, "and you can tell the tenor of the times." It was not a tenor he liked. He laid the blame for the erosion of local customs at the feet of the wealthy and powerful, whose wasteful spending on rites, courtesies, and clothing was driving everyone beneath them into a frenzy of conspicuous consumption that was washing away the core Confucian values of decorum, modesty, and concern for the moral welfare of others. As he specifically contrasts the wealthy and powerful with the gentry, we know whom he was targeting: the great commercial families of Songjiang. Their wealth was bringing sweeping changes to life on the Yangzi delta, which Gu Qing then patiently enumerated one by one for his readers: twenty-three changes, to be exact.[53]

We needn't recite all twenty-three changes to see what was troubling Gu Qing; a few will suffice. The presents that the families of brides and grooms exchanged before the wedding, for example, had hugely escalated in value. Funerals had become unnecessarily elaborate and prolonged. The little gifts that used to lubricate social intercourse had grown into large bribes. Dinner parties had moved from a modest table of vegetables and fruits to a groaning board of meat and fish laid out on expensive porcelain. The unadorned four-sided hat that the Ming founder had mandated as male headgear had given way to elaborate hats, to say nothing of the absurd concoctions that women's headdresses had become. Simple cloth shoes had been replaced by fancy embroidered footwear. The curtains on sedan chairs had changed; so too had the design of pleasure boats. Ordinary stationery had disappeared in favor of gilt-edged letter paper. Even dye colors had changed. Now it was lychee red instead of peach red. Kingfisher blue had gone out in favor of sky blue. Incense brown had pushed aside soy-sauce brown. And so forth. Gu's list of transgressions concludes with the (to him) shocking revelation that the rich were dressing the young male actors they hired for their private theater troupes in purple gauze outfits. "Observe the changes and you can tell the tenor of the times."

For every one of his twenty-three indictments of local extravagance, Gu Qing employs the same grammatical construction: "originally" people did this, and "now" or "recently" they have started to do that. Neither "recently" nor "now" was to be regarded as the way things should be. To his credit, Gu steers clear of phrasing his objection to this conspicuous consumption as a moral affront, a complaint that would be much heard toward the end of the century. Matters were not that far gone in 1512. He simply objected to the wasteful stupidity of spending good money on nothing more than "eye-catching decoration." Gilding the lily added nothing of value to the lily, but did spell bankruptcy for the family that felt it had to keep up with the Wangs. And the burden of the game fell hardest on those least able to afford the candle.

The motor of this consumption, though Gu does not phrase it in this way, was simple enough: the growth of supply and demand for luxury goods. There had been a time when peach red, kingfisher blue, and soy-sauce brown were the only shades one could hope to buy in those colors. By 1512, dye producers had made it possible to switch to lychee red, sky blue, and incense brown, all presumably offered at much higher prices. The same escalation was true on the consumers' end. There had been a time—"originally"—when most people would have been unable to even think of buying anything fancier than the soy-sauce shade when they needed brown dye. "Now" eager consumers could afford to move up to incense brown, want to be seen doing it, and be pleased about the whole business. If the changes Gu Qing disparaged had inundated consumption practices, it was simply because people could now afford to join fashion's flood. Gu interpreted these changes as signs of creeping decadence; we, on the other hand, can take them as clear evidence of the new prosperity into which the Ming economy had grown by the turn of the sixteenth century.

This new prosperity has led some historians to argue that the aggregate wealth of the Chinese economy meant a high standard of living for individual Chinese. Adriano de las Cortes, a Spanish Jesuit who was shipwrecked on the coast of Guangdong in 1625, was not convinced. Las Cortes was impressed with the productivity of the economy, but he distinguished that productivity from the prosperity of ordinary people. "The volume of merchandise that the Chinese possess is not a sufficient argument to prove that they are very rich," he observes. "Speaking in general, this is a people that, on the contrary, is extremely poor."[54] Las Cortes was on a remote coastal frontier of Guangdong, not in the grand cities of Jiangnan. He made his judgment relative to the wealth and poverty of places he knew from home. From his experience on both sides of the globe, the rural people of the Ming were no better off than the rural people of Europe, possibly worse. Either way, the difference was probably slight at the bottom of society, where most people had just enough to get by.

Trees and Tigers

Despite Las Cortes' caution, the economy of the Ming was cumulatively more prosperous than at any earlier time in Chinese history. Growth may have been modest by our exaggerated modern standards, but it drove up the amount of food that a larger population was consuming. As it did, economic growth pressed on natural resources. To increase acreage under the plow, farmers created polders by walling in and draining the margins of lakes and rivers. Inland, they constructed terraced fields up steep hillsides. These alterations increased the production of grain needed to support growth in the other sectors of the economy, but they were not without significant cost. Polders raised water levels and intensified the danger of flooding. Terracing reduced the natural habitat of plant and animal species and leached the thin soil off the hills. Deforestation and drainage to construct Chinese-style agricultural systems in upland zones resulted in a massive reduction of the biota that had been essential resources enabling upland people to reproduce themselves and their society.[55]

People in most times and places—and especially in urban places and modern times—have assumed the natural world to be larger and more abundant than we now do. The wealthy dilettante Zhang Dai could join a winter hunting party in the hills outside Nanjing in 1638 and return with one deer, three musk deer, three pheasants, four rabbits, and seven foxes, and not show any awareness of the impact this recreational cull might have had on the natural ecology of the zone into which the hunters went for their amusement.[56]

The depletion of animal populations was less noticeable, though, than the thinning of the more slowly renewable resource of trees.[57] China had not been heavily forested for a good millennium. Already a half-millennium before that, the philosopher Mencius explained the process by which evil conditions turn people to evil by describing the stripping of wood from Ox Mountain, leaving it eroded and infertile.[58] Every Yuan and Ming schoolboy knew the passage, which means that every schoolboy was aware of the negative effects of clear-cutting. It does not mean that every schoolboy thought it would happen to his particular mountain.

The demand for wood not only denuded the hills in the heavily populated zones of eastern China but caused timber to be stripped from areas ever more distant from centers of population. Once the hills around the North China Plain lost their last stands, many blamed the ambitious palace-building of both dynasties. In his blueprint for administrative reform submitted to the throne in 1487, Qiu Jun advised that wood requisitioning for palace construction in North Zhili be reduced and that procurement be spread more equitably across the country. He also advocated an integrated policy of borderland reforestation. As the forest historian Nicholas Menzies has noted, such recommendations permitted some forest recovery, yet "prohibitions on felling proved to be an unsatisfactory way to protect forests" when local authorities had other ideas or the central state faced emergency needs.[59]

One option was to source timber from ever more distant locations. This the Ming did by cutting its way through the upland forests of Yunnan province in the far southwest. By 1537, an imperial censor wrote to the emperor from Yunnan warning him that "in every location timber and wood oil have been harvested in such volume that these areas are becoming exhausted" and advising that the government reduce its consumption of wood. He was concerned less with the trees than with the increasing burden this placed on local labor, but his concern enables us to see ours.[60]

The polymath Tan Qian, whose *Miscellaneous Offerings from Date Grove (Zaolin zazu)* is possibly the largest commonplace book of the seventeenth century, includes in it a curious monograph recording this loss. *Ancient Trees (Gu mu)* documents every old tree for which he could track down some record, organized by province and county. He opens on a somber note: "In the heavily trafficked parts of the realm, of the trees of a circumference so great that several must join hands to encircle them, not one in ten has survived the carpenter's axe. In the obscure depths of mountain valleys and on the twisting slopes of shadowed hills, trees may be able to accumulate their years, yet even there people can't help but measure them up, so that they are unable to escape misfortune." He then complains that the sources for compiling a catalogue of ancient trees are woefully incomplete, especially in places where recent cutting has been severe. "As much of the terrain of Shaanxi, Sichuan, Fujian, Guangxi, Yunnan, and Guizhou is difficult to access, lumberjacks over the centuries have stripped out excellent timber" without anyone noticing or leaving any record of what has been taken.[61] His distress is evenly distributed between the missing entries and the missing trees.

No part of China was more thoroughly deforested by the end of the Ming than South Zhili. From the entire Yangzi delta north to the drainage basin of the Huai River, not one tree of any age or worth makes it into Tan's catalogue. The other province that stands out for the thinness of its forests is Guangdong, which was becoming an increasingly populous and ecologically strained zone in the Ming. The only provinces where good stands of old-growth timber still remained were Shanxi and Shaanxi in the northwest and Yunnan in the southwest. Spirit Forest, a stand of over a thousand ancient trees in northern Shaanxi, managed to escape destruction because it was believed to be haunted. Where such taboos were absent, old-growth stands were cut down. Only in Yunnan were there still trees of sufficient girth to provide the pillars and beams for temple and palace halls. Tan's monograph suggests that the historian Mark Elvin may have been too optimistic in concluding that "China's general forest crisis is only about three hundred years old, even if in a few areas, such as the lower Yangzi valley, its roots are considerably deeper in time."[62] Even if shortage was not yet strangling the economy, *Ancient Trees* suggests that the crisis was just over the horizon. Densely populated zones had no significant forest resources by the end of the Ming, and the western periphery was losing its forests at a pace too rapid to chronicle.

The cutting of the forests caused more than the trees to disappear. It also hastened the disappearance of animal habitats. The disappearing forest animal that attracted the greatest attention from contemporary commentators was the tiger. The tiger is at the top of the food chain—what Robert Marks has called the "star species" of south China. A few tigers may still survive in the hills along China's southern border; in the Yuan and Ming they could be found from south China to Siberia, though their numbers diminished as farmers encroached on agriculturally marginal areas to feed ever more mouths. The more they did so, the more tiger and human habitats overlapped. Tigers need up to 100 square kilometers of unspoiled terrain to survive. As such natural tracts dwindled, tiger spotting became something of an obsession for Ming writers—not as rare as spotting a dragon, but worth recording every time it happened.[63]

The earliest losses were in the north. The *History of the Ming* records an encounter between a tiger and the literary traveler Qiao Yu (js. 1484) at the top of Hua Mountain in Shaanxi province. What is surprising, besides that it occurred so far north, is how the encounter turned out. Qiao's servants threw themselves on the ground in terror, but Qiao simply sat down facing the tiger and remained motionless. The tiger dropped his tail and skulked away.[64] The compilers of a dynastic history did not deliver homilies, but they expected readers to catch the implication: Qiao was a man of such moral strength that hostile nature could not unleash its destructive forces against him. The tiger obeyed. The trope became popular: tigers cast in the role of base nature defeated by the power of the human mind.

Not all encounters with tigers ended so peaceably. South of the Yangzi, Huizhou prefecture was still heavily forested in the Ming. It was also infamous for its tigers, which regularly entered inhabited areas and attacked people. Local sources, when speaking of the tigers, use the language of "poison," "harm," "catastrophe," and "disaster." In 1410, a county magistrate ordered 314 tiger traps dug. Within a month, forty-six tigers had fallen into these pits and been killed. Their elimination was celebrated as a positive step toward domesticating the wild. But the tigers of Huizhou refused to disappear. By 1600 they had bounced back so strongly that a later magistrate launched a second eradication campaign against them.[65]

Minister of Rites Huo Tao (1487–1540), famously intolerant of popular religious practices, brought his Confucian spiritual resources to bear on the tiger problem while en route from Jiangxi to his home in Guangzhou. Halfway through Guangdong province, he stopped in Qingyuan county. Residents told him they were being plagued by tigers along the river valley and asked him to do something about it. Rather than organize an eradication program, Huo opted for a ritual solution. He submitted a statement to the spirit of the local mountain demanding that he restrain the tigers. As a result, so the account goes, the tigers disappeared.[66] Huo's success was simply a matter of timing, for he was calling on the spirit of the mountain to banish the tigers just past the environmental tipping point between human and tiger ecosystems.

Buddhists tried a similar approach to the tiger problem. Chan (Zen) master Zhiheng had founded a monastery outside Hangzhou in 967 on Yunqi Mountain, a natural habitat of tigers. Known as the Tiger Tamer Master, he was said to have kept the tigers in line by feeding them meat he bought for the purpose. The monastery was destroyed by a flood in 1494. When the Buddhist monk Lianchi Zhuhong (1535–1615) revived it in 1571, the tigers were still there.[67] Zhuhong approached the tiger problem through Buddhist logic. Violent creatures are reincarnations of people who, having caused and suffered much violence in their past lives, are reborn to act out that karma. Tigers were troubled spirits, or "hungry ghosts," working through their reincarnation. Soothe these spirits, and the problem would go away.

Zhuhong's rituals to ease their karmic burden failed to have any effect, so in November 1596 he conducted a massive five-day pacification ceremony. "I believe that human beings and tigers originally possess the same nature, and the cause for the destruction lies in hatred inherited from the past," he explained in a text he wrote to commemorate the rite. "If we capture the tigers, then we harm one another. If we drive them away, then what is the difference between us and other people? Thus we must perform fasting and create merit so that we may hope to transform them silently and the harm will quietly disappear." He enlisted spiritual help akin to the sort Huo Tao invoked, asking "all the saints who had tamed tigers since ancient times . . . to carry this prayer to the gods of the mountains and the earth in all directions." Acknowledging the need to find a balance between humans and tigers, he "beseeched those who had harmed the lives of tigers in their previous lives to renounce their anger and resentment, so that the tigers would not seek retribution." Tigers who killed people could not help themselves, he explained, but humans could, and should, if they hoped to "cultivate the heart of compassion." Zhuhong ended his text praying that the tigers "will speedily live out their present incarnations and depart from the wheel of suffering."[68]

When Zhuhong revived the monastery in 1571, the region was suffering drought. The local farmers, who had encroached onto Yunqi Mountain and built fields here, had asked him to pray for rain. When he brought rain, he was rewarded with the support he needed to rebuild the monastery. Herein lies the best way to eradicate tigers: convert their wilderness habitat to agricultural land. As this happened, tigers lost their capacity to resist human encroachment. By the end of the Ming, tigers were confined to the southern provinces; by the end of the eighteenth century, they had almost disappeared.[69] Poachers killed and ate what may have been China's last wild tiger in 2009.

Population, commercialization, and the expansion of the Ming economy drove people to strip resources from nature at a greater rate than ever before, turning wilderness into fields and hunting larger mammals to near-extinction.[70] As the economy grew, habitats disappeared, forests shrank, and the human relationship to the environment became ever more fragile. In 1642, one of the last ancient trees on the Yangzi delta, a 300-year-old tree on the grounds of Zhu Yuanzhang's tomb outside Nanjing, was cut down and its root dug out for firewood. When the Ming fell two years later, many believed that this desecration had brought the dynasty down.[71] Perhaps they were right.

6

FAMILIES

LI GUANGHUA died on September 21, 1612, in the Yangzi River town of Shashi, Sand Spit Market, midway between the Sichuan Basin and the East China Sea. It was not the outcome he intended. He had shown promise as a young student, and by the age of twenty-three had won a coveted spot as a *shengyuan* or licentiate in the county school, the first step to examination success and bureaucratic office. But the success he dreamed of eluded him. Every time he sat for the exams, he found himself below the cut-off of those who passed. While Guanghua was pursuing his studies, his younger brother Guangchun had gone into business up the Yangzi River. The success that eluded the elder brother came to the younger. As Li Guanghua found himself with a growing family to support—four sons plus an unknown number of daughters—he decided to give up study and join his younger brother's business at Sand Spit Market. It was there he died at the age of forty-seven.

When the news of Guanghua's death reached his family, his eldest surviving son, Li Hua, set out for Sand Spit Market to bring the body home. By the time Hua delivered his father's body back home on March 6, 1613, his tomb on the ancestral burial hill was ready. A geomancer or *fengshui* master skilled in the art of reading the lines of energy that flowed through the landscape had been brought in to find a site that would ensure success in business for his heirs, so the family was prepared to bury Li Guanghua the following day according to the rites appropriate for a man who had attained the exalted status of *shengyuan*. The day after the burial, his gravestone was set up in a ceremony at which, according to the text inscribed on the stone, his three filial sons wept tears of blood.

The story of Li Guanghua, as of all such ordinary people, would normally have ended there in the spring of 1613: the aspirant to official rank who went into business and did well enough until death found him in his prime. His tablet would have been placed in the lineage shrine to receive the regular sacrifices given to the ancestors. He would have been remembered for a few generations within his own family, his name would have been preserved in the family shrine a few generations more, and then he would have been forgotten, except perhaps in the pages of his lineage's genealogy. As it happens, his gravestone prevented his story from ending there, for the stone was swept into a commodity market such as Li would never have imagined. It surfaced in the wholesale antiques market in Seoul, Korea, and from there found its way into a furniture store in Toronto to be sold as a garden ornament. I rescued it there in 2002, and it now resides in my office in Vancouver.

The stone is not a fine piece of work. Cut from a sheet of slate, it bears an inscription in a rough calligraphy carved into the stone surface with minimal skill. The weather has worn the text, rendering some of it indecipherable, but enough remains to reconstruct Li Guanghua's life of modest success. The one striking element are the two characters at the top of the stone, the "sun" on the right and "moon" on the left, the very characters that are combined to create the word *ming*, "bright," the name of the dynasty. Sun and moon mark the grave to situate the deceased within the cosmos and claim for him their protection in death—just as they did for the emperor in life. (The left shoulder of his court robe was emblazoned with a red sun and his right with a white moon; see Fig. 2).[1] Emperors got all the protection and notice they wanted. Li Guanghua has only this gravestone, all the more precious for that reason. The stone does not tell us who Li Guanghua really *was,* but it does show us what defined his life: the kin among whom he lived and died.

The Kinship Matrix

The people of the Yuan and Ming lived within an administrative matrix over which they had little control, but they also lived with one of their own making, the matrix of kinship. Your identity and status depended far less on the state than on whom you were related to. Your father was your first important asset. He produced or acquired the food that kept you fed and accumulated or lost property that you inherited. He was also, as importantly, the first link in the chains of brothers and male cousins that integrated everyone into the kinship networks that anthropologists call lineages. An agnatic lineage was the group to which you were tied through your father and that bore your surname. With these men and their families, you shared descent from a common ancestor, along with the ritual identity that this sharing provided. It was the network where males could look for land and capital in good times, for relief and protection in bad, and for burial and sacrifice in worst.

The lineage, however sharply bounded, was not closed, nor could be. Out around it extended an unstable and usually growing set of relationships with other lineages via the women who married in and out. Agnatic ties endowed you with your root identity, but affinity—kinship by marriage—connected you to the world beyond your doors. Your affines gave you spouses, neighbors, friends, and trading partners. They were the channels between the silos of agnatic lineages. They were so important that if a son died before marrying, his parents might arrange a posthumous marriage with a family that had recently lost a daughter and hold a full wedding for the deceased couple on the day after burial, rather than lose the chance to establish an affinal connection.[2]

Not all of these practices originated during the Yuan and Ming dynasties, but many did. This was a time when the social nature of families was changing. The old aristocratic families of the Tang were gone, and the court families of the Song were disappearing. It is rare to find a prominent family in the Ming that is able to claim any sort of elevated ancestry extending back before the Yuan. Elite families still emerged in Yuan and Ming society, but they faded much more rapidly than the great families of earlier times. A county magistrate in Fujian, for instance, notes approvingly around 1572 that one can still find some "old families of the previous dynasty" in the wealthier southern part of the county, but not in great numbers, and nowhere else in the county.[3]

To compensate for the more flattened social space they inhabited, families sought strength by organizing themselves into larger kinship networks that shared resources. The most successful lineages owned agricultural land, stocked granaries, provided graveyards, built ancestral shrines, and ran businesses. Some set up primary schools for lineage children, and many provided financial sponsorship for the brightest boys so that they could study with tutors in preparation for the state examinations. Lineages kept careful records of their members and assets, and increasingly through the Ming published a selection of their records in lineage genealogies that were shared with potential business or marriage partners.[4]

Li Guanghua's epitaph reflects elements of this structure. Unlike a genealogy, which accumulates data over time, an epitaph captures the state of a family at one moment. Li's is quite informative if read carefully. It supplies the personal names of his grandfather and father, who define his agnatic descent. Of his mother we have only her surname, Zhang, for like most women her husband's family has not preserved her given name in their record. Li was the eldest of three brothers. Li Guangchun, the second, brought Guanghua to Sand Spit Market. The third was Li Guanghuan, who married but died young. The epitaph also refers to a sister. So the nuclear family into which Li was born consisted of a couple, three sons, and one daughter, with two sons and the daughter surviving to adulthood.

Li Guanghua was able to improve this record of reproduction within his own family. His first wife, Madam Zhou, died without issue, but his second wife, Madam Zou, gave birth to four sons. The first, Yao, did not survive infancy but the next three—Hua, Qing, and Xuan—made it to adulthood. There was also an unmarried daughter or daughters at the time of Guanghua's death. The epitaph reveals that Madam Zou was from Township 11, which implies that Guanghua was not. Elite families might well seek to extend their territorial reach by intermarrying across their county, but ordinary people found their partners locally.

Families used names to post relationships and distinguish generations. The names of Li's four sons were identified as siblings by using characters that shared a script component or "radical" (the "fire" radical in this case). This was one common naming device; another was to give all the children in one generation the same middle name, which is what Li's father did with Guanghua, Guangchun, and Guanghuan. Either of these naming devices could also indicate birth order, for sometimes names were chosen that formed a sequence. Another common practice within a lineage was to give a label to each generation, and assign a number to each cousin according to seniority. Unusually for a text of this sort, the epitaph specifies everyone's position in the lineage this way. Li and his brothers were in the *zheng* generation: Guanghua was *zheng* 2, Guangchun *zheng* 8, and Guanghuan *zheng* 9. Li's sons belonged to the *rui* generation. His eldest son was *rui* 3, indicating that one or two men in the *zheng* generation, probably first cousins, had already sired second cousins before Li Yao was born. The only other fact about the lineage in the epitaph is the existence of what it terms an "ancestral mound," presumably the lineage cemetery, though this part of the stone is so badly eroded that the full text cannot be reconstructed.

Li's epitaph also identifies some affinal connections. The surname of Zhou figures prominently. Guanghua, Guangchun, and a sister all married members of the Zhou family. Likely they were all from the same lineage. Marriage ties were carefully structured, being critical to family survival. Only Guanghuan married into another family, the Lis (not the same Li), though Madam Li was remarried to a Xie after Guanghuan died. In the next generation, the wife of Li Guanghua's first son, Hua, was a Chen. After Hua died without an heir, the Lis married Madam Chen off to a Zhou male. A Zhou nephew, the son of Guanghua's sister, is also mentioned in the epitaph. The repeating marriage connection between the Lis and the Zhous suggests that both lineages were linked to each other and securely embedded in their local society.

The Lives of Women

If kinship provided the matrix of life for the people of the Yuan and Ming, gender was its organizing principle. Gender imposed hierarchy by arranging social relations in ways that favored men over women. Female infanticide is a convenient example: families that had to keep their size small killed girls before boys rather than give up either male labor power in the future or sacrificial continuity, which was thought to pass through males, in the even further future. But gender could also impose complementarity.[5] Women were as necessary to the reproduction of the family, in all senses, as were men. The organization of family life took this into consideration by gendering labor, sending the men out into the fields and keeping the women at home to weave. The actual division of labor within the family could vary from this norm. In the north, for example, women were expected to contribute to agricultural labor by handling some aspects of grain processing, such as milling (Fig. 10). The growth of the commercial economy had the capacity to upset the gendered division of labor. When silk was woven to sell on the market rather than supply domestic needs, for instance, men stepped in to take over the women's work, and the income. Only when silk was industrialized at the end of the Qing dynasty did they return the work to the women.[6]

Fig. 10 An old woman and a young woman husking millet using a stone roller. The inscription names four types of millet and explains that "this roller is used for all of them." The elegant stools in the foreground may have been intended to present their work as domestic rather than agricultural labor. Their difference in age is indicated by the wrinkled face of the woman on the left and the elegant coiffure of the young woman on the right, presumably her daughter-in-law. Song Yingxing, *The Creation of Things by Heaven and Artifice* (*Tiangong kaiwu*), 1637.

We know about the women of the Yuan and Ming mostly from the records of their marriages. Marriage placed an asymmetrical burden on women. They had to move between families after the marriage while men did not; they could marry only one spouse while men could take secondary wives; and they could not remarry once widowed whereas men could (as Li Guanghua did when his first wife, Madam Zhou, died). These, at least, were the expectations. In the matter of widowhood, the state provided an incentive to women who remained "chaste widows" after their husband's death in order to honor them as moral exemplars. A woman who was widowed before the age of twenty-nine and remained unmarried until she passed the age of forty-nine qualified for an official citation and a banner proclaiming her chastity. This honor was significant, the only one the state conferred on women. The husband's lineage filed the chastity claim and applied for certification, as this accomplishment redounded to the credit of the entire lineage for supporting her in this noble endeavor—even if, as was so often the case, the support was grudging, and the reward was hers.[7]

Widow remarriage was far more common than widow chastity. Many women married more than once, as we should expect in a population in which female infanticide meant that men outnumbered women and the odds of a husband dying while the wife was still in her child-bearing years were high. Traditional morality objected. The early-Ming philosopher Cao Duan (1376–1447) advised that a widow suspected of being involved in a liaison with another man should be given a knife and a rope and locked in a cowshed until she committed suicide.[8] In practice, it was uncommon for a widow to remain unmarried. The deceased husband's family put pressure on a widowed woman of child-bearing age not to hang around as chaste widows but to remarry within five years at the outside, lest the cost of supporting her become too great a burden on her husband's lineage, or she be forced by poverty into unchaste ways.[9] Li Guanghua's epitaph testifies to this practice. When his brother Guanghuan died young, Madam Li was married out to a Xie. And when his son Hua died, Madam Chen was married to a Zhou.

A widow with a son had some chance of resisting remarriage, if she so chose, since her husband's lineage should have an interest in not compromising his line of descent. Still, there was no guarantee that her husband's relatives would leave her with the resources to do so.[10] A woman named Qiu Miaozhen in early-fifteenth-century Yangzhou married a Huang and found herself widowed at age twenty-six with a young son. She chose not to remarry, and her husband's lineage did not try to dispossess her of the use of her deceased husband's property. However, a brother-in-law had his eye on her property and pressed her to remarry so that he could take possession. Qiu was able to muster her in-laws against him by performing a public libation in their presence vowing never to abandon her dead husband. According to her biography, she was triply rewarded. She attained the great age of eighty-nine, had a grandson who passed the highest exam in 1484 and rose to be a vice-minister, and received the honor of being addressed as *shuren,* "woman of virtue," a title normally reserved for wives of officials of the third rank.[11] Qiu Miaozhen was one of the lucky ones. In bad times, women who chose widow chastity might have no option other than suicide.[12]

Biographies of chaste widows preserved in local gazetteers often record age of marriage, status and age of husband, age at widowhood, number of children, and length of widowhood, as these data were necessary to file a chastity claim.[13] They indicate that women married between the ages of fifteen and nineteen, though most were married by the age of seventeen. Cao Duan of the knife-in-the-cowshed advice regarded thirteen or fourteen as the age at which marriages could be contemplated.[14] It was not unheard of for a girl to marry even as young as twelve. This was the absolute lower limit, as Yuan law treated the sexual penetration of girls under the age of twelve as rape, even if the act was consensual.[15] These statistics are roughly confirmed by the compiler of a 1530 gazetteer from coastal Fujian, who observes that girls were expected to marry between the ages of thirteen and nineteen, and explains that marriage after nineteen increased the risk of birth complications.[16] Marriage also had its north-south difference, with women in the south and the interior marrying roughly a year earlier than women in the north.

A woman in the Ming on average gave birth to four children who survived infancy, though it was rare for more than two or three to reach adulthood.[17] The pressure to produce a male heir was intense, increasing the number of life-threatening childbirths that women had to face. The diarist Li Rihua reflects on this reality when he records the death of his daughter-in-law, Madam Chen, on August 20, 1610. She suffered an attack of what Li calls "womb fever," and so a doctor was called in. After ten doses of the medicine he prescribed, the fever abated. She went into labor early, and the birth was easy enough that she was able to get up and oversee the washing of her newborn daughter. After eating some thin gruel, however, she fainted. Li rushed into the women's quarters to see if there was anything he could do, but she was already dead. Married at sixteen, she was dead at eighteen.[18]

Childbirth interested Li, and a year later he notes in his diary two local cases of quintuplets. One of the mothers and all her babies died, but the other mother and babies survived. Li was baffled to understand how a woman could bear this many children at once. "Twins are not so many as to be considered strange, but having five babies is almost the same as dogs or swine. For there to be two cases in one county—might that not be some sort of auspicious sign?"[19]

The one way in which a woman could opt out of marrying and bearing children was to become a Buddhist nun. It was not a common recourse, given the huge pressure on women to perpetuate male lines. Besides, Confucian prejudice placed women who chose monastic celibacy under suspicion of sexual promiscuity. Minister of Rites Huo Tao (1487–1540), who banished tigers from Qingyuan county, was particularly virulent on this topic. In 1537 he submitted to the emperor a long memorial alleging gross sexual misconduct by the nuns in Nanjing. "Without husbands or family, without father or mother, without children to care for: aren't they pitiful? By name what they do is cultivation; in fact they are destroying morality. They sully themselves, and they sully other men's wives as well" by pimping for women who came to the temples to pray. "Aren't these women horrible?" The Jiajing emperor agreed and approved Huo's proposal to expropriate seventy-eight convents in the region for other public uses such as schools or shrines to figures honored by the state. Nuns over the age of fifty were to be sent back to their families or assigned to homes for the elderly. Women below that age were given three months to find a husband. If they failed, they would be assigned as wives to unmarried soldiers.[20]

A satirical writer of the early Qing suggests that the Buddhists later evened the score. Huo, it was said, had his eye on a choice monastic property that he wanted to make into his own private residence, and so he included it in the expropriation order. The last monk to leave penned this on a wall: "A scholar's family has moved into a home for monks. Does this mean that now his wife will be lying in an old monk's bedroom?" The moral counterstrike worked, and Huo was shamed into abandoning his plan.[21] Moral attacks tended to dominate public discourse, but the real issue, as everyone understood, was property. Buddhist nuns and monks lived in institutions that controlled property; and in an economy in which land was highly valued, their communities were always under threat from competing landlords. Sex was the furthest thing from anyone's mind, except perhaps the accuser's.

Besides wife or nun, there were other paths for women, notably concubine and prostitute. The demand for both services was high, and only increased with the commercialization of the economy and the accumulation of the wealth required to pay for them. Concubinage was a legal form of polygamy to which wealthy men resorted in most cases because of failure to produce a male heir. It was an expensive undertaking and tended to create instability within the family, as primary wives—who could be divorced for failing to produce a boy—feared the loss not just of affection but of status and the share of property that went along with status.[22] The idea of multiple marriage partners was a popular male fantasy that runs through the novelistic literature of the late Ming, most famously in the erotic novel *Plum in the Golden Vase (Jinping mei),* which recounts the dissolute life of a wealthy northern merchant.[23]

A wider spectrum of men subscribed to prostitution, especially the poorest and the richest. The poorest had no prospect of ever affording the gifts that had to be sent to a bride's family nor the property that was needed to set up an independent household. Engaging a prostitute along one of the alleys that every town had for such services was their sole source of sexual satisfaction. The richest moved in a different circle of high-class brothels and skilled entertainers, some of whom might provide sexual services and some of whom might not. It could be a sordid world for the women who worked in it, though there are cases of women who, when presented with the choice to return to their natal families, preferred to stay in their business.[24] There are also famous instances, though far fewer, of entertainers who achieved a level of education and cultural sophistication that won them immense admiration among the male elite of Jiangnan. If high-class concubinage was eroticized, high-class entertainment was romanticized in tales of equal emotional partnerships between talented entertainers and lofty young men.[25] It was a nice idea, rarely achieved.

The Lives of Men

The Ming world was a different place for men. Male dominance was practiced through the ritual and social superiority of men over women. But this ritual superiority was also the soft underbelly of male dominance, for it forced families to construct themselves in peculiar ways.

First of all, a family had to have a son. Without a son, the parents and the father's ancestors could no longer receive sacrifices, for only male heirs could tend the ancestors' spirits. Every society finds ways around its own rules, and Chinese devised ways to surmount this problem. The lineage could transfer the son of a brother or cousin into the line of an uncle without a male heir; or the family could adopt a daughter's husband into the family, a practice known as uxorilocal marriage; or a childless man, if he were a devout Buddhist, could endow a Buddhist monastery to perform rites in perpetuity for himself and his ancestors.

Where the burden of male ritual superiority fell heaviest was on girls. A family forced to reduce its size in the face of disasters or financial difficulties still had to ensure the continuity of the family line, and that entailed sacrificing its female children, either by selling them or killing them. Female infanticide was punishable under Ming law, but that was not a disincentive strong enough to prevent the practice, and most magistrates turned a blind eye. We can guess at the scale of the practice from population data, and these suggest that the scale was huge. However cooked and unreliable the figures were, they consistently point to a gender imbalance starting around 90 females for every 100 males and sinking as low as 50 or even less.[26] Part of this imbalance may be a statistical mirage, but the fact is that large numbers of women who should have been there in a normal population were missing.

The immediate cost of population control by postpartum termination must be measured in the deaths of females, but there was a delayed cost for men, and that was imposed celibacy. There just were not enough women to go around. This situation led to some ingenious arrangements. One was what anthropologists call fraternal polyandry: the practice of brothers marrying one woman. A village on the Zhejiang coast nicknamed Handkerchief Gulch became notorious for just this practice. The village got the name from a local custom. When one of the brothers wanted to sleep with their wife, he hung his handkerchief by her door, alerting his brothers to stay away. Poor women were said to like the arrangement, as it promised greater financial security than being married to just one income earner. Some claimed that the Japanese introduced this practice. That allegation may simply be a matter of projecting nonstandard practices onto foreigners; it could also have arisen because of the high risk associated with seagoing, and in this way have become associated with Japanese mariners. The custom was banned in 1491 by analogy to Article 392 of the Ming Code against fornication with relatives by marriage.[27]

Wife-sharing may have been suppressed in Handkerchief Gulch, but other arrangements were devised to provide sexual satisfaction for the males who were prevented from marrying. One of the less common was marriage between men, which was practiced in parts of Guangdong and Fujian. The standard character for male *(nan)* consisted of the sign for "field" *(tian)* on top and "strength" *(li)* on bottom. For men who entered such a marriage with another man, "strength" was replaced with "female" *(nü)*. Again, the custom was associated with seafaring, on the understanding that men who were isolated from women for long periods of time resorted to each other for sexual satisfaction, though the shortage of women must have played a part in regularizing this type of relationship.[28]

The majority of men who entered heterosexual marriages tended to do so at a later age than their wives by as much as five years.[29] For a husband to take a wife older than him was a practice largely confined to the north. A Fujian writer in 1530 gives the ideal age of marriage for boys as between fifteen and twenty-four, significantly higher than the range for girls.[30] This practice may be yet another effect of the shortage of females. The competition of numbers, combined with the obligation to pay a bride-price to the woman's family, delayed their entry into marriage. For some, the delay became permanent. We will never know how many men did not marry, as the unmarried were less likely to be documented than those who were members of families, but their share of the male population may have run as high as 20 percent.

The household of Fu Ben displays many of the features that characterized marriage in the Yuan and Ming, as well as a few of the exceptions. Fu was a simple farmer who would have remained unknown to us were it not that Tan Qian copied his Yellow Register certificate of 1398 into his commonplace book, *Date Grove Miscellany.*[31] Fu Ben farmed in the Yellow River valley in central Henan province, a location to which he had migrated not too long prior to the date of his registration there. The certificate records that Fu's household owned a three-room house with a tiled roof and 200 *mu* of hill land. In 1398 the family consisted of three generations. The eldest generation consisted of Fu Ben (fifty-one years old) and his wife (forty-one). The second generation included their son Chou'er (nineteen), his wife (unnamed, twenty-two), and two unmarried daughters, Jingshuang (twelve) and Zhaode (eight). Chou'er and wife had started a third generation in the person of a son who was as yet too young to be formally named. He was registered under his nickname, Little Sledgehammer. Little Sledgehammer is listed as their second son, indicating that an earlier son had not survived infancy. From this registration, we can see that Fu Ben delayed his marriage, probably because of poverty, possibly because of the difficult conditions early in the Ming, possibly because he had been a young soldier at the time of the dynastic transition. In the next generation, both Chou'er and his wife married in their late teens. Fu Ben was ten years senior to his wife, whereas Chou'er was three years younger than his, and had already fathered two sons by the age of nineteen. Finally there are the two Fu daughters. Jingshuang was still only twelve, too young to be sent into marriage, and Zhaode another four years younger. It is worth noting that the Fu family had managed to raise two daughters in the wake of a son without sacrificing either of the girls. In this family, Little Sledgehammer was the prize member. He would see the Fus into their fourth generation and maintain the sacrifices for the other three generations once all had passed away.

With these responsibilities came the obligation of maintaining the family not just ritually but financially. The great burden of males in the Yuan and Ming was to produce food and wealth sufficient to keep the family alive. Fu did so by accumulating 200 *mu* (12½ hectares) of land. This apparently sizeable acreage was registered as hill land, the fiscal category for the least productive agricultural land: Fu was not a wealthy man. Still, it was enough for him and Chou'er to support their household of seven members. It would be the obligation of Chou'er and Little Sledgehammer as the males in the household to attempt each in his turn to increase the family's wealth.

Occupational Households

When the Mongols conquered China, their instinct was to freeze the social order in perpetuity into a four-level ethnic structure. They were at the top. Next were the People of Various Categories (*semuren,* commonly mistranslated as Colored-Eyed People), those who were neither Mongol nor Chinese. In third position came the people living in north China, tagged as Han People *(hanren)*, a term derived not from a memory of the Han dynasty over a millennium in the past, but from the Han state, one of the Sixteen Kingdoms that controlled the North China Plain in the fourth century.[32] The lowest category was for the Southerners *(nanren)*, basically the former subjects of the Southern Song, the people whom the Mongols trusted least.

Within this structure the Mongols introduced a system of fixed occupations for men to freeze labor and ensure the production of the goods and services they required. Whatever occupation someone had been pursuing at the time of conquest, he should continue in that occupation. Bow makers, for example, were tagged forever as hereditary bow makers. Their households were permanently registered as such, and their sons required by law to continue in their fathers' work, part of which involved manufacturing bows for the Mongols. This system of household registration was an elaborate simplification that captured the population at one moment and sought to hold it in place. The Mongols were not bad sociologists; they just wanted to make sure that the manufacturing economy over which they presided would always furnish them with what they commanded. At base, the concern was fiscal, not social: to make sure that they could levy the goods and labor they required. They had no particular interest in how a household constituted itself as a "family."

No complete list was ever drawn up at the time of every category that Mongol administrators identified. Huang Qinglian in the 1970s attempted to round up all the references he could find to household designations so as to map out the administrative matrix of Yuan households. He came up with a list of eighty-three. A few designations are ethnic but most are occupational. The first four are the largest: civilians, soldiers, artisans, and couriers. Many categories had subcategories. Soldier households, for example, were subdivided into twelve groups, such as gunners and archers. Further down the list appear some fairly specialized households, such as ginger-growers. There were also categories for what might be described as religious professionals. Confucians, to their considerable dismay, found themselves down here among Daoist priests, Buddhist monks, Buddhist nuns, and devotees of the Buddhist redeemer Maitreya.[33]

Zhu Yuanzhang when he came to power had a similar urge for clarity, but his simplification took on a different character. Being the great unifier, he had no use for the Mongols' ethnic structure. The great majority of his subjects were "Chinese" in any case. For his fourfold structure, he turned to classical authority, declaring that he, like the ancient sages, would "rule the realm by dividing the people into four occupations of gentry, peasant, artisan, and merchant. When all four are properly distributed throughout the realm, there is never a shortfall in meeting the needs of the nation."[34]

These four were not occupations so much as broad categories, and early Ming officials had the good sense to leave them underspecified. They allowed the fine occupational distinctions of the Mongols to fall into disuse, except for a few holdovers for certain specialized tasks, such as salterns (salt makers). By midway through the dynasty, however, hiring was regarded as a more cost-effective way of producing salt than obliging specially registered saltern households to provide the labor on a hereditary basis. Despite the social fluidity that came to characterize Ming society, the ancient model of the four categories of the people continued to enjoy an ideological status out of proportion to social reality. Anyone who lamented the sorry state into which the world had sunk had only to invoke the fourfold classification and feel that a healthy alternative from the past was still available, even though it wasn't.

The one category of occupational household that lingered was soldiers. The Ming required soldiers, and had at least a million men in arms at any one time.[35] Soldiering was not a lucrative career, but it did have a security that other professions lacked. More importantly, it earned the soldier a state salary that did not expire with the man's death. According to rules laid down in the Hongwu reign, his widow continued to receive it, albeit at a reduced rate, as a pension. A son born within the first ten years of his father's service as a soldier could inherit his father's post and salary, so long as he passed the test in military skills. Even if the son suffered from a physical incapacity that prevented him from soldiering, he could still receive his father's stipend at a reduced rate.[36] This system of payments failed in practice to keep military manpower up to the level that was required, for increasingly military skills became devalued and soldiering families turned to more lucrative careers, even studying for the exams. The Ming state was determined to block the exodus from the military, such that even if someone from a military household passed the Presented Scholar exam, his household would not be excused from that status. Only someone who rose to the post of Minister of War was permitted to petition for a change of status.[37] In practice, though, the registration status into which one was born was not a sufficient barrier for men of ambition to escape military service.

Of the four status categories, the one that made the Hongwu emperor most uneasy was the gentry. He knew he needed their literacy and learning to administer the affairs of seventy million people. But he also suspected they would always place their own advantage ahead of the needs of himself and be a burden on the people. As the economy grew, the fourfold order shifted, pushing farmers to the bottom of the heap and sometimes even pushing the gentry below the new heroes of the age, merchants.

Gentry Society

The gentry—those families who groomed their sons for service in the state bureaucracy and supported themselves through landowning for the most part—emerged as an increasingly coherent local elite in the fifteenth century to dominate local society through a range of economic and ritual practices. Their most important resource as a class was their access to the examination system. In theory, any boy could sit for the exams; in practice, passing the exams depended on being able to afford the intensive literary education needed to master the texts and writing styles that the examiners tested, as well as a cultural comprehension of what the entire system of texts represented. The system was a two-edged sword, of course, for just as the exams were the entryway to gentry status, so too they posed the major threat to that status. Gentry families could perpetuate themselves as such only by returning to the exams successfully at least once every other generation, and for many this proved an impossible task.

The very possibility of an examination-based elite at the local level was defeated in the Yuan by the simple fact that the court held few examinations. The Mongols preferred to bypass anything so autonomous and unpredictable as exams and simply appoint those whom they trusted. When Zhu Yuanzhang came to the throne, he was not entirely sure he wanted the gentry to constitute itself as a social class, hence his foot-dragging over the reinstatement of the examination system. By the end of his reign, however, the exams were back in place, and it was just a matter of time until a group of families emerged to dominate local society as they dominated the examination system.

There were other routes into state service, but none that carried the status of success in the national examinations. The competition was enormous, however, and only grew over time as the number of candidates grew. By 1630, for example, the provincial examination compound in Nanjing, one of the largest in the country, had 7,500 cells. If every cell was filled, that meant that only one candidate in fifteen could go on to the national Presented Scholar examinations in Beijing the following year.[38]

The pinnacle of the system was the Grand Secretariat, the small group of four or five men who advised the emperor and oversaw all court business. One of them was Shang Lu (1414–1486). The inscription on his burial portrait lists his accumulation of official titles: "Grand Secretary of the Palace of Respectful Caution and concurrent Minister of Personnel holding the honorary title of Senior Tutor to the Heir Apparent, Optimus in all three examinations leading to the rank of Presented Scholar, and posthumously honored with the title of Cultured and Resolute"—which is a long-winded way of saying that he was the highest civil official in the realm (Fig. 11). Between 1479 and 1544, only twelve other men became Grand Secretary of the Palace of Respectful Caution and concurrent Minister of Personnel holding the honorary title of Senior Tutor to the Heir Apparent.[39] Of the twelve, two missed out on the posthumous honor of receiving the title of Cultured. One of these was the infamous Yan Song (1480–1565), whose monopoly over power in the Jiajing court resulted in his disgrace and the confiscation of all his property at the age of eighty-two. Though well regarded, Shang Lu did not make a mark as a grand secretary, holding that post for only two years. His greatest claim to fame was ranking first in the provincial, national, and palace examinations—the only person ever to do so in the Ming.[40]

Few who entered the exam system would ever get near the post of grand secretary. The bureaucracy was too large, and there were too many trying to enter the system. According to the author of a late-Ming commonplace book, at the turn of the sixteenth century there were 20,400 civil officials, 10,000 military officials, and 35,800 students on government stipends.[41] If we assume conservatively that every county had roughly 150 licenciates registered at the official Confucian school, and multiply that figure by the number of counties, there would have been well over 150,000 young men at the bottom of the system clamoring to move up. Add to these the far more enormous body of students vying for spots in the Confucian school, and the field of competition was enormous.

The examinations had effects other than turning out bureaucrats. They put young scholars into contact with others of the same social background and ambition at ever higher levels all over the country. Sitting for the provincial exam involved more than the isolating exercise of mastering a common body of knowledge, showing up, writing your answers in your sealed cell, and then going home. It was a highly social experience that involved sharing accommodations with other aspirants often for weeks at a time, eating and drinking together, sometimes forging deep bonds. If you passed, those who passed with you became your cohort with whom you could expect to associate, and for whom you could be called upon to do favors, for the rest of your life.

Participating in the system could also be a linguistically transforming

Fig. 11 Portrait of Shang Lu (1414–1486). The inscription reads: "Funerary portrait of Master Shang, Grand Secretary of the Palace of Respectful Caution and concurrent Minister of Personnel holding the honorary title of Senior Tutor to the Heir Apparent, Optimus in all three examinations leading to the rank of Presented Scholar, and posthumously honored with the title of Cultured and Resolute." Arthur M. Sackler Museum, Harvard University.

experience. Students were required to learn "official speech" in their home schools, an artificial pronunciation that Europeans called Mandarin, following the Portuguese adaptation of the Hindu word *mantrū* (Sanskrit *mantrim*) meaning "counselor." Mandarin tended to echo Nanjing pronunciation, as one would expect of a regime based in that region, though it acquired a literary lexicon that floated above any actual dialect. The transition to fluency in Mandarin increased the higher one rose through the system, though some officials never escaped their dialects. In 1527, before he became Minister of Rites, Huo Tao was excused from his duty as Court Lecturer to the Jiajing emperor on the grounds that he was a southerner and "his speech was full of incomprehensible words."[42] This failing was not enough to cripple a career, but it was enough for teasing someone who, like Huo Tao, famously lacked a sense of humor. Fujian accents were judged to be even more incomprehensible. Two Fujianese served as grand secretaries in the 1430s, but according to the *History of the Ming,* "their speech was so hard to understand that two hundred years passed" before an emperor appointed anyone from Fujian to the Grand Secretariat.[43]

Given the impossible odds, most educated young men eventually abandoned the examination mill and sought more local ways of achieving wealth and status. One was to become a gentleman farmer. Another was to go into business, as Li Guanghua did. A third was to take up a profession, among which the most honored was medicine. Doctoring had become a major alternative for educated men during the Yuan, when examinations were rarely held and the gentry turned their energies to other pursuits. The social status of doctors rose accordingly.[44] The writer Wang Daokun, who was from a prosperous mercantile lineage in Huizhou and happily praised the virtues of merchants, also appreciated good doctors. He boasted that the greatest doctors of his day were all from Huizhou. "In my prefecture we revere doctors as much as we revere Confucian scholars," he declared. Good doctors were all said to have started out in Confucian studies before taking up medicine. A Huizhou doctor named Wu Shanfu explained to Wang that the method used to train Confucian scholars is directly transferable: "Confucian scholars start with mastering the skills of canonical interpretation, and then proceed to the Hundred Schools, and on this basis construct their skills. Doctors must proceed in the same way," starting with the canonical texts and then working their way through the works of the four leading physicians of the Jin and Yuan dynasty known as the Four Masters. Wu felt that the ancient texts and the Four Masters yielded treatments for at most four cases out of ten, however. For the other six cases, the physician must turn to the Way in order to find a cure. "The Way is here," he declares, pointing to himself. "When I need to make up a prescription," he says, "I go to the Way."[45]

Whatever they ended up doing, the local gentry were concerned first and foremost with protecting their elite status in local society. This meant deploying wealth and social connections, securing ties to the local magistrate, and engaging in charity and patronage so as to enhance their public image. The family that was successful in managing such enterprises could still earn a respected position in local society without actually sending a son through the examination system, though eventually a son, or a son-in-law, would have to return to that font of status. These conditions meant that the gentry could never become an aristocracy, in the sense of an elite who were born to that status by virtue of their birth. In the Ming, the only aristocracy were the members of the Zhu family who could prove descent from the founder, and they lived in a sequestered world of privileged isolation. But gentry families did not simply rise and fall with each generation. They enjoyed reputations, connections, and wealth that enabled them to coast for many generations, constituting what might be called an aristogeny, a system in which the elite reproduced itself across generations without relying solely on birth.[46] The secure presence of local gentry in every county throughout the country from the fifteenth century forward testifies to their success (Fig. 12). A family could survive for two or three generations without a member gaining a degree, so long as it maintained social and marital ties with the families who did and continued to participate in the shared activities that distinguished the elite from ordinary families. Still, an examination title was in the end the real marker of elite status, and every family that could afford the education put their sons to study.

Beneath the gentry, according to the traditional model of the "four categories," were the other three: farmers, artisans, and merchants. Farmers were praised in official ideology as second only to the gentry. The logic was obvious, for their labor ensured that grain was grown, and grain furnished the foundation of the country. In the language of agrarian political economy, agriculture was the root of the polity, just as it was the root of the economy. All other enterprises, whether in manufacturing or trade, were what that language called the "branches." They might grow luxuriantly, but only when the root was firm. Or that was the ideology.

Fig. 12 Portrait of a Ming gentleman. His elegant Tang-style hat signals that he was a member of the gentry, though the absence of any insignia of title or rank suggests that he was not himself a degree-holder. Arthur M. Sackler Museum, Harvard University.

Throughout the Yuan and Ming, the status of farmer was in real terms the lowest in the society. The status of artisans remained constrained until the last legal limits to free labor were removed in the sixteenth century, though by the Wanli era, as we shall later see, some were happily hobnobbing with the gentry.

The group whose real status became most divorced from their official status was merchants. This category included anyone from the smallest of rural peddlers to the great families who earned vast wealth through state monopolies, most particularly salt. Zhu Yuanzhang's insistence on reinstating the four categories may have arisen in part because of popular resentment of the enormous wealth and power that merchants were able to achieve under the Mongols at the expense of other occupations. Local records contain not a few stories of arrogant merchants in the Yuan dynasty who get their comeuppance. Indeed, the earliest unofficial dragon sighting I have found for the Yuan dynasty, in 1292, tells of a wealthy man crossing a river near Lake Tai on the Yangzi delta. At mid-crossing, the boat ran aground, and the boatman's pole got stuck trying to force it free. When the man ordered a servant into the water to try lifting the boat, the poor man discovered that the boat was stuck on the spine of a dragon and the pole lodged in its scales. The merchant panicked and leapt overboard, but was unable to swim. He ordered one of his attendants to get him ashore to safety. Upon returning home, however, they all fell ill and the merchant died.[47] It was, in the popular mind, a fitting end to a rich man.

Merchant wealth could be a stepping stone to official status, but that usually involved infiltrating a gentry family through intermarriage or patronage. Writing late in the fifteenth century at a time when commerce was on the rise, Lu Rong notes in his *Miscellany from Bean Garden* that "today's wealthy families that have risen from a base condition always attach themselves to esteemed lineages" and will do anything to entangle younger members of these lineages in order to gain a foothold in the upper class. One device was to "combine genealogies," which involved grafting one's family onto the lineage of a gentry family and taking that surname. The Yangzi delta became notorious for this practice. Lu gives the example of Kong Kerang, a descendant of Confucius in the fifty-fifth generation. Kong's grandfather had served the Yuan as a tax overseer, but being of a scholarly bent, had shifted into teaching. This was a sure recipe for poverty, into which the family duly sank. A family of rich merchants in the neighboring prefecture spotted an opportunity and set its sights on seducing Kong Kerang. He adamantly refused, but his family was in such dire straits that eventually he allowed the merchant family to "trade a boatload of rice for the genealogy." "Many are the descendants of the sages who are deceived by the lowly bent on fooling the world," Lu concludes.[48]

Formal notions of status still prevailed, but increasingly the substance inside the form was simply money, and commerce was the sphere in which such money was to be made. By the end of the Ming, the barrier between gentry families and merchant families had ebbed to an all-time low. A wealthy family's strategy for long-term success was to field sons in both ventures: the one to build the family's wealth, the other to enhance its status.

Family, Propriety, Property

This portrait of Yuan and Ming society has dwelt on the administrative and kinship matrices within which families struggled either to survive or to better themselves. What drove these strategies was the unwavering goal of reproducing the family into the next generation. Women's lives as wives and mothers were shaped by this goal, and their status within the family depended on their ability to deliver an heir. The lives of men, whether as students, farmers, or merchants, were shaped by it as well. They had to have sufficient wealth or status to impregnate a wife or concubine, and they had to pass that wealth and status down to their sons to do the same.

Society recognized four rituals cementing the kinship ideal of continuity of the male line. These had been distilled by the Song founder of Neo-Confucianism, Zhu Xi (1130–1200), in a handbook entitled *The Family Rituals (Jiali)*. Not until the mid-Ming did this book begin to circulate extensively, first among the gentry who sought to purify their practices and thereby elevate themselves above the rabble, but eventually among everyone. By early in the eighteenth century, according to a Jesuit missionary, it was second only to the Confucian *Analects* as the book most often found in a Chinese home.[49] The first ritual was capping, which marked a boy's passage into puberty and therefore his capacity to reproduce. Though this rite was viewed as archaic, some gentry families revived its practice. The second was the wedding, which gave the boy the female he needed to reproduce. The third was the funeral, which served as a great occasion for celebrating the ritual unity of the lineage. The fourth was ancestor worship, which recognized the position of the deceased within an ancestral line and provided him with the comfort of sacrifice without which he would suffer in the afterlife as a hungry ghost.

These rituals made present and reinforced the dominance of the patriline. But under them lay the foundation of all families, property. The closing story in this chapter illustrates how ritual and property needs overlapped, and the lengths to which some men at the lower edge of elite status would go to preserve both. It involves a legal dispute that reached the Hongzhi emperor on November 29, 1499.[50] This was unusual, as almost all legal cases that did not involve homicide were resolved at the county level by the magistrate. This one had become far too complicated to stay local.

Wang Zhen owned a plot of land in the hills outside the city of Nanchang, the capital of Jiangxi province. In this hilly region south of the Yangzi River, population was dense, land scarce, and the locals often on the move elsewhere looking for work or land. "The hills are many and the fields few," as a Nanchang county author noted by way of explaining why the local people were so lean.[51] Even the hills that were used for graves rather than fields were at a premium. The most coveted bits of upland topography were those spots where professional *fengshui* masters judged that the lines of energy *(qi)* streaming through the landscape converged. Bury an ancestor within such an energy field and the deceased's spirit will bestow fortune to his descendants. Jiangxi lineages competed for the best tomb sites and resorted to tricks and violence in their struggle to improve their fortunes at others' expense. Grave land feuds were endemic to the province through the Ming and Qing dynasties.

The case arose because another man, Zhang Yingqi, buried a body in Wang Zhen's plot without his permission. Zhang was a student on stipend at the Nanchang government school. An aspirant to elite status, he appears to have been from a lesser gentry family, though the surviving case summary in the court digest, the *Veritable Records of the Hongzhi Reign,* reveals little about either man.[52] Wang Zhen, the owner, was not a student, nor did he possess any token of gentry status. Yet even a commoner could take his case to court, and this is what Wang did, filing a lawsuit with the prefecture. Jiangxi natives were notorious for their litigiousness; indeed, the Hongwu emperor in his final instructions to the people in 1398 singles out Jiangxi in this regard, complaining that they "cannot endure even minor matters, and go directly to the capital to bring suits."[53] Down to the end of the Ming, according to a local commentator, the people of Nanchang were "avid in work and stingy in giving, indifferent to duty and happy to quarrel; cunning and glib, litigious and libelous."[54] Wang Zhen was among his own.

The prefectural judge agreed to hear Wang Zhen's case and decided in his favor. Zhang decided to go on the offensive. He turned to a fellow student, Liu Ximeng, for something he needed: a connection to a higher authority. Liu had a job tutoring the son of an assistant provincial surveillance commissioner, Wu Qiong. Liu had managed to ingratiate himself with the Wu family by handing out presents, thereby positioning himself as a small-time power broker between the commissioner and anyone who might want his help. The connection suited Zhang's purpose, for an assistant surveillance commissioner outranked a prefectural judge. Money changed hands, and a word from Liu in someone's ear in the Wu household succeeded in getting Wang's case against Zhang overturned.

Zhang did not rest content with this victory. As the offending corpse was still in the ground, Zhang was still vulnerable to counterattack from Wang, so he filed a lawsuit against him, this time turning to another patron, Assistant Education Intendant Su Kui (js. 1487). Su had a reputation for refusing to act on behalf of private interests, which makes his willingness to help Zhang, allegedly after receiving a bribe, puzzling. Zhang must have misled him into thinking that he was truly the injured party. Meanwhile, seeing that Zhang had support elsewhere in the bureaucracy, Wang Zhen looked for a connection to another reservoir of state power that had a local presence, the eunuchs of the imperial household. How he got to Grand Defender Dong Rang, a eunuch whom the emperor had sent out to supervise regional security, is not known.[55] Presumably the right sum of money could open any door, so long as one knew which corridor to travel. Wang presented his case to Dong, and Dong obliged him by having both Zhang and Liu thrown in prison, where torturers could persuade them to withdraw Zhang's claim.

Dong intended only to intimidate the two students into backing off, but the torturers got more than Dong meant them to. The hapless pair revealed that they had bribed state officials to back their side of the lawsuit. Once this revelation came out, what had been a local property dispute turned into a bureaucratic crime. It now had to travel up the administrative hierarchy to Beijing, first to the Censorate, then to the Ministry of Justice, and finally to the throne. Dong Rang had gone too far, and now the emperor was looking down at the situation in Nanchang and demanding that the ministry investigate. The modest capillary action of rational bribery from below would now be reversed by the percolating gravity of state authority from above.

The investigation uncovered that eunuch Dong had already marked Su Kui as his enemy before Wang Zhen ever came to him for assistance. The rivalry between them is what gave the bribes their traction. Dong was chafing under a perceived insult from Su on another matter and for that reason had agreed to take Wang Zhen's side. It turns out that Su was not alone in disliking Dong, whose high-handed activities on behalf of the imperial household had already goaded other officials into petitioning for the man's removal, without effect.[56] Dong was in the stronger position and managed to get Su thrown in prison on a corruption charge. Students at the Nanchang school were so offended by the eunuch's attack that a hundred of them stormed the jail and freed their superior. Su was exonerated and later promoted, yet Dong was left untouched.[57]

The Hongzhi emperor chose to stay out of their squabble, reasoning that no actual damage had been inflicted on any of the parties. He reprimanded Dong Rang and Su Kui for adjudicating lawsuits they were not entitled by their positions to entertain, and he reprimanded Su Kui and Wu Qiong for taking payoffs. The burden of his judgment fell away from his officials, however, and landed most heavily on the two students who started the affair. Zhang Yingqi and his friend Liu Ximeng were stripped of their studentships and stipends. This punishment might appear light, but it was harsh in a status environment as competitive as the mid-Ming, for it banished them permanently from becoming gentry.

As for the buried ancestor, Hongzhi said nothing. The incident was too far down the bureaucratic structure for him to be able to see clearly what had happened. As he admitted that January in another edict, "We live deep within the Nine Walls [the palace], and though We stretch Our thoughts over the entire realm, there are places Our ears and eyes do not reach and where Our grace has not been manifested."[58] This was a matter of scale, but it was also a matter of what should concern an emperor. Corruption within his bureaucracy mattered greatly, for if he could not rely on his officials to report impartially about conditions throughout the realm, he could not rule justly. As for where Zhang's ancestor was interred, that was not his concern. It was a dispute best left to the prefectural court.

At this point, the case goes cold. We do not know whether Zhang Yingqi had to disinter his ancestor and bury him somewhere else. The negative judgment from the throne must have weighed heavily against him down in the prefecture, so it is reasonable to think that he had to move the body, and that Wang Zhen got his land back. What is interesting, and gets to the heart of the rules by which Ming society worked, is that in the divergence between Zhang's and Wang's interests, property prevailed. Zhang's concern was that his ancestor be buried in such a way that his spirit could rain positive benefits on his descendants. Wang undoubtedly wanted the same benefits for himself, but his more immediate concern in the case was that someone had tried to take control of his property.

Ritual propriety mattered to the state, which had to rely on the individual, the family, and society at large to observe the proprieties that kept all the moving parts in place. In practice, however, all an emperor could do was single out exceptional instances and intervene. Had this case come before the founding emperor, one can imagine him executing everyone involved. Fortunately for Zhang Yingqi, the Hongzhi emperor was not one to smash a small target with a large hammer. What mattered in any case was the sanctity of private property, without which none of these systems could function. In a lawsuit over a grave, the sensible path was to let propriety defer to property, not force property to rearrange itself around propriety. Rites may have held families together, but only property enabled them to survive. Wang Zhen had to win, and did.

7

BELIEFS

THE people of the Yuan and Ming believed that the cosmos consisted of three powers or realms. Heaven was above, Earth was below, and they were in between. Heaven was the creative power that oversaw everything, but at a great distance: only the emperor, its son, could pray directly to so august a power. But Heaven was also a realm that thronged with gods, to whom Daoist priests, Buddhist monks, and in fact anyone could pray. Guanyin, a favorite protector of women and children, was one of the gods of the Buddhists (see Fig. 5). Below Heaven lay Earth. This was the realm where Humans lived. So too did the lesser deities like the Stove God and the Door God, puckish spirits who meddled in the everyday lives of Humans. And so too did the spirits of the ancestors, to whom regular sacrifices had to be made lest they feel neglected and make their descendants miserable. But the Earth was not only the surface that Humans plowed or dug down into. Deep within Earth lay the Earth Prison, a vast purgatory where the deceased were consigned for 27 months to be purged of their sins before King Yama and the rest of the Ten Gods who oversaw the prison released them for their next reincarnation.

It was the lot of Humans to live between the forces of Heaven and Earth. Over the centuries, Chinese had developed three sets of beliefs, institutions, and liturgical methods to deal with that predicament, or more precisely, how to live a good life: the Three Teachings of Daoism, Buddhism, and Confucianism. Daoism, which honored the patriarch Laozi, offered naturalistic technologies such as charms, spells, and remedies to help them adapt to the physical conditions of the surrounding world. Geomancy—the siting of buildings in relation to the forces that flowed through the Earth's surface—was one of their technologies. Buddhism provided a rich corpus of ideas and institutions that offered release from the attachments that caused suffering. It was also the religion that handled death, its monks and monasteries supplying ritual services to help the living assist the dead in navigating their passage through the Earth Prison.

Confucianism, a loose set of doctrines derived from the teachings of Confucius, proposed a different course. Confucians held that the path to goodness was more likely to be found by moral training and the effort of relating ethically to others than by going through the gods. During the Song, Confucianism had undergone an enhancement into what was known as the Teaching of the Way, which we now call Neo-Confucianism. This enhancement had profound philosophical effects, some of which did not come fully to fruition until the Ming, but it had no real impact on the beliefs that guided ordinary people through their lives. For them, the Three Teachings were all they needed.

After invading, the Mongols found themselves on a more complicated religious terrain than they were used to. Their religious orientation would best be described as shamanism. This led them to an interest in Tibetan Buddhism (Lamaism) more than in the Three Teachings, yet they feared the gods sufficiently to show respect toward any religious professional who could claim to have access to esoteric knowledge. Buddhist monks and Daoist priests competed for Mongol patronage, debating each other most famously in a grand court debate in 1258, but the regime declined to support one and suppress the other, so long as adherents of each could demonstrate the value of their teaching for bringing good fortune.[1] It may have been the effort of Chinese trying to explain the multiplicity of their traditions to the Mongols—or was it the Mongols trying to reflect their understanding back to their new subjects?—that encouraged the people of the Yuan to see an underlying unity in their systems of beliefs. The Three Teachings did not inhabit three separate categories, but rested within a single category. Sectarian adherents of the three different traditions still fought amongst themselves for state recognition and preferential treatment, but those who were less partisan in their beliefs felt free to engage eclectically, even erratically, with all the traditions.

The unity of the Three Teachings—the idea that China's three main belief traditions were no more than different expressions of the same thing—is one of the two big ideas that shaped the landscape of belief during the Yuan and Ming dynasties.[2] The second big idea is one that Confucians associated with the mid-Ming statesman, Wang Yangming, also known as Wang Shouren (1472–1529). We met Wang in Chapter 4 as the commander of the campaign against indigenous insurgents along the Guangxi border, and who refused to join the top elite of the day in condemning the Jiajing emperor for desiring to honor his natural father and mother above the emperor he succeeded. His new idea was that moral knowledge was innate and not learned. This discovery led to the radical belief that while we should study the writings of the sages, we should do it to reveal to ourselves what we already know intuitively. Among his more radical interpreters, however, there grew up the idea that studying the Classics was less useful for acquiring moral knowledge than engaging in meditation and moral introspection. Replacing study with intuition was a stiff challenge to traditional pedagogy, which regarded pupils as ignorant until taught.

Both ideas—that we are hard-wired to think and act morally, and that the Three Teachings are essentially one—may have been shaped by a deep desire for the one thing that the dismantling of the Song world had rendered impossible: unity. The conviction in the importance of unity was congruent with the times. Just as the new Yuan regime had taken up unity as its first political principle, a principle entirely seconded by the Ming, so too their subjects turned to unity as a principle to guide both religious and philosophical thought. Unity reconnected what experience had forced apart, made whole what had been broken. And it had been there from the very start. As we shall see, come the Ming, even some Neo-Confucians willingly engaged in Buddhist practices and experienced no fundamental contradiction between ethical practice in one sphere and religious practice in another. The notion that all teachings shared an underlying unity became a template for dealing with yet another teaching that arrived in the wake of European trade, Christianity.

Entering the realm of Yuan-Ming thought through the concept of "belief" rather than "religion" or "philosophy" may be somewhat unconventional, but I do so in order to capture how people approached their understanding of the world. I do so not to dismiss what they thought was true as merely a matter of belief. Truth was of course what they thought it to be, but they were well able to distinguish truth from belief when they found contradictions between them. Belief held the same range of meaning for them as it does for us: religious devotion, personal trust, the reliability of well-established facts. They, like us, relied on belief; so too they understood that belief was subject to challenge and change. Take for instance Xie Zhaozhe's complaint in his commonplace book, *Five Offerings,* that the story of Daoist patriarch Laozi spending eighty-one years in the womb is "certainly not worth believing."[3] We share Xie's belief on this point, but he embraced many convictions that we would doubt. Whether what they believed is what we consider real and true is beside the point, which is that some questioned the prevailing ideas of the time as much as others believed in them. As we are about to see, there was much to disagree over: what happened to you after you died, what determined the nature of the physical world, whether the Earth was flat, and what constituted a moral life: these were matters over which people could and did disagree. Through the sixteenth and seventeenth centuries in particular, inquiring minds were constantly examining the world, consulting books, revising their assumptions, and wondering how best to make sense of what they experienced. Our task is not to reduce their beliefs to ours, but see how they worked.

Rather than starting with the philosophers, we will begin with the man who was uninhibited in expressing his most deep-seated beliefs, as well as in imposing them on others, the Hongwu emperor.

The Spirits

At dawn on the morning of February 17, 1372, the Hongwu emperor led a long procession of officials in full court regalia out of the Forbidden City in Nanjing, heading for Jiang's Hill Monastery in the hills five kilometers to the east. Jiang's Hill had been Nanjing's premier Buddhist institution in the Yuan. The sixth Yuan emperor had visited it early in 1325, offering a donation that inspired vast public giving for a rebuilding project. Five years later his nephew continued the imperial patronage by presenting the abbot with a ceremonial robe. Zhu Yuanzhang followed suit. Both before and after he became the Hongwu emperor, he visited Jiang's Hill many times. But this time was different. The purpose of this visit was to attend a plenary mass for the souls of all who had died in the civil war that had brought him to power.

Presiding at the mass was the eminent Chan (Zen) Buddhist master Qingjun (1328–1392; Fig. 13). As the imperial procession approached, Qingjun led a thousand monks carrying flowers and incense to meet the emperor at the front gate of the monastery. Hongwu presented Qingjun with ten thousand ounces of silver, an extraordinary gift, then was ushered into the Great Buddha Hall. Facing north—a striking gesture, as emperors always faced south except when praying to Heaven—he led his of-

Fig. 13 A Buddhist monk in the guise of a *lohan* (arhat), one who has achieved nirvana through spiritual cultivation and will not be reborn in this life. Carved in the last half-century of the Yuan dynasty, this wooden statue was originally painted. Royal Ontario Museum, Toronto.

ficials in bowing before the statue of the Buddha as an orchestra struck up "Doing Good in the World," the first of seven hymns they would play through the day. The assembly bowed again, and then Hongwu knelt—another remarkable gesture, as emperors knelt only to Heaven—and presented offerings. A troupe of twenty-eight monks bearing lamps, jewels, and lotus blossoms danced before Hongwu, after which he knelt and made offerings again. After more hymns and dances, the emperor and his officials filed outside. Hongwu mounted a throne platform prepared for him fifty paces inside the front gate and sat facing south to receive prostrations. At dusk, the emperor led the procession back to the Forbidden City.

As the next day dawned, snowflakes fell and an auspicious glow filled the sky. Everyone gazed at Heaven with awe—or so we are told by the account of this ceremony by Song Lian (1310–1381). Song was Hongwu's senior Confucian advisor, the architect of his regime, and a devout Buddhist himself. His report testified that the emperor had succeeded in exciting the grace of the Buddha, that Heaven had given him its miraculous response.[4]

It is obvious what people at the time were to make of this: they were to agree that it had indeed happened and to be in awe of the new emperor's cosmic power. And he *was* new. Hongwu was just starting his fifth year as emperor. These were early days for a dynasty, when anything could happen to derail the fragile new state. Military force, administrative rigor, and Confucian moralism had been put to use to create and sustain the new regime, but there was another weapon in the arsenal of state-making—religion—and Hongwu and his advisors were now using it. As the art historian Patricia Berger describes another imperial Buddhist service in 1407 at the same monastery, the heavenly portents accompanying the ceremonies "were harmoniously and diplomatically observed by all, in an extended moment of 'consensual hallucination.'"[5] If the new dynasty did not have Heaven's blessing, no one was prepared to argue the matter.

Hongwu used Buddhism for this plenary mass because it was the religious tradition for dealing with death. He makes no reference to Buddhist doctrine in the edict he promulgated ordering everyone in the realm to offer obeisance to the Buddha on the day of the mass—though he does mention his stint in a monastery when he was an orphaned teenager. What he talks about instead was the popular notion that beyond this, the World of Light in which we live, lay the World of Shadow, where the unshriven souls of the dead exist as hungry ghosts. Confucianism scorned this idea of a haunted afterlife, as it rejected the entire theology of the Earth Prison. At death a person divided into his aerial aspect, his *hun,* and his earthly aspect, his *po.* The *po* sank to Earth where it quickly degraded, while the *hun* ascended to Heaven where it dispersed into the air. Of the original person, nothing substantial or spiritual remained. This was a rigorous view of the afterlife, and was not accepted by most people. They preferred to believe in the ongoing existence of the dead, whether in purgatory or as lost souls wandering just beyond the edge of sight in this world.

So did Hongwu. It was essential to his understanding of why people chose to be good. Without spirits to placate, morality had no goal and a plenary mass made no sense. Hongwu admits in the edict announcing his intention to parade to Jiang's Hill that the purpose of the plenary mass was to ease the spirits of the many innocent people who died during the recent civil war, and to acknowledge the living who still suffered under the burdens that the war placed on them. It was the memory of this war that Hongwu was struggling to redress. He understood the scale of injustice and the need for redress, and turned to Buddhism so that something might be done.[6]

Most of Hongwu's Confucian advisors would have disapproved. He illustrates their skeptical view of the spirits in an essay he wrote in response to a report he received of spectral beings who flickered between the World of Light and the World of Shadow. After reading the report, Hongwu turned to his advisors and asked them what they thought. The easy course for them would have been to avoid Hongwu's beliefs and parrot Confucius' two pieces of advice on this point to his disciples: that they should respect the spirits but keep them at a distance, and should not concern themselves with whether the spirits were really present at a sacrifice but conduct the rites as though they were there.[7] But one attending official took the bait.

"This is utter nonsense," he spluttered. It was the reply Hongwu was hoping for. "How do you know?" he retorted. Rather than turn immediately to Confucius' view, the courtier launched into a brief lecture on the nature of physical human existence to disprove the reality of the phenomena that were reported.

> People are born possessing the material *[qi]* of Heaven and Earth. This is why physically they go from young to mature, mature to se-

nior, senior to senile, and senile to dead. At the moment of death, the *hun* rises to Heaven and the *po* sinks to Earth. The *hun* is made of spiritual substance but is exhausted when it reaches the empyrean and is blown to the four winds. The *po* on the other hand consists of the bones, flesh, and hair. When it touches the Earth, it decays into dust and mixes into the mud. That is why Confucius refused to talk about the spirits.

Hongwu believed differently. His beliefs came from his impoverished childhood in the Huai River valley. They stayed with him because they described his understanding of where good and evil came from. Failing to respect the spirits was what had led to his world being full of corpses and hungry ghosts. "If one believes there are no spirits," he pointedly reminded his courtier, "then there is nothing to fear in Heaven or Earth and nothing with which to nourish the ancestors—what kind of person would think this?" The difference in beliefs between Hongwu and his courtier ran right through society—between the educated Confucian elite on one side and the majority of ordinary people, among whom Hongwu counted himself, on the other. For them, the living and the dead inhabited worlds that were largely the same, and so close to each other that sometimes they intersected.

A simple demonstration of this conviction is a land purchase contract that happens to survive from November 1361. Ye Fengshu bought the plot of land from Wu Wangyi, though a careful second reading of the contract reveals the astonishing (to us, at least) fact that Ye Fengshu was dead at the time of the purchase, and the land he was buying was his own grave plot. The contract is signed not by Ye but by his wife, Li Dingdu, on his behalf. Following her signature, as well as the signatures of Wu Wangyi and the agent, there appears this four-line vow:

> Whosoever signs this, the cranes that ascend to Heaven [shall know],
> Whosoever signs this, the fish that swim through the water [shall know];
> Should the white cranes read this, they will carry [what they know] to purest Heaven;
> Should the fish read this, they will take [what they know] to the ocean's depths.[8]

This is not standard contract language. Li Dingdu added it to make sure that the agreements she made on her husband's behalf in the World of Light would be honored in the World of Shadow. There were courts in purgatory, and so there was every reason to think that, as in the real world, someone might want to dispute the sale. Accordingly, she buried a copy of the contract—written on pottery so that it would not decay, as paper would—in the grave so that Ye would have access to it to prove that the plot had been bought and paid for. If a posthumous legal challenge got ugly, the final arbiter would be the boss among the Ten Kings in the underworld, King Yama. Ye might well need to show him proof.

The belief that the dead went to a purgatory was not restricted to the common people. Even the incorruptible Chief Grand Secretary Ye Xianggao (1559–1627) admitted to Chief Censor Zou Yuanbiao (1551–1624) that he was susceptible to the idea that he would be punished in purgatory were he to stray from upright conduct.

"You, sir," Ye said to Zou, "talk about Confucius, whereas I talk only about King Yama." "Why?" Zou asked. "Well," he replied, "I am worthless and old, and the day when my mortal remains will fill a bunghole is nigh. If I cheat my sovereign or lead the country astray, harm people or damage assets, seduce the powerful or accept bribes, I will have to watch my deeds tallied before the throne of King Yama—which is why I don't dare to do any of these things."[9]

Zou laughed, as Ye meant him to—but Ye was only half-joking, as we know from the keen interest he showed when Jesuit missionaries told him about hell in their religion. Ye had no difficulty believing simultaneously in the teachings of Confucius and in purgatory. Zou did, but then both men lived along the spectrum of skepticism and belief that characterized their age.

Buddhism and Daoism

The fact of the matter is that most people of the Yuan and Ming were promiscuous in their beliefs. They were able to accept that Buddhism, Daoism and Confucianism were different modes for grasping the same reality. But not all the Three Teachings were equal. If we were to think of belief as a multi-story structure, then Ye Xianggao's conversation shows that popular Buddhism occupied most of the ground floor. The Buddha's existence as a supreme deity was widely believed, and almost as widely accepted, though some strict Confucians demurred. What went on higher up in the edifice depended on personal choice and individual experience. Ordinary people prayed to Buddha's many images and burned incense in hope of gaining his attention and favor. At times, some of the gentry were drawn into his worship, gathering at the great monasteries that dotted the scenic uplands throughout the realm, drawing inspiration as much from the scenery as from the sutras.

This orientation encouraged fourteenth-century intellectuals such as Tao Zongyi to experiment with the conceptual structures of the Three Teachings, trying to match each with the other. He depicts his sense of their unity most vividly in his commonplace book, *Notes after the Plowing Is Done (Chuogeng lu)*, first published in 1366. Toward the end of the book, Tao sketches three diagrams, one for each Teaching, to display the relationship among the eight principal concepts of that Teaching, and in such a way as to show them to be homologies of one another. The point of the exercise is clear: despite their intricate differences, the Three Teachings express an underlying unity.

The Hongwu emperor shared this view when he declared the legitimacy of all three at the beginning of his reign, but he favored Buddhism. His early posture as a patron of Buddhist monks and monasteries may have derived from his own experience as a teenager who threw himself on the mercy of a monastery in order to stay alive after losing his parents to famine. He was careful, however, not to adopt Buddhism as a state religion and not to allow himself to be drawn too far into Buddhist interests. After the Hu Weiyong purge, Hongwu suffered a crisis of confidence about the system he had put in place. No longer was he willing to trust anyone, and that included trusting the Buddhist monks to whom he had played the role of royal patron. After 1380 he suspected Buddhists of harboring the same selfish designs as everyone else—pursuing personal gratification, evading taxes, and perhaps plotting sedition. To bring them under tight control, he imposed a series of tough regulations. The low point came in 1391. The Hundred-Day Edict announced that monasteries throughout the country were to be amalgamated into a tiny handful in each county, and a permanent ban was imposed on founding any new ones. These changes had to be carried out within a hundred days. The restrictions were completed three years later with the Edict of Seclusion, which ordered monks to stay inside their monasteries and not to mingle with the public.[10]

These restrictions were relaxed as soon as the Hongwu emperor was dead, but the suppression lasted long enough that many of the monasteries never revived. Later emperors might choose to be patrons of Buddhism or Daoism (Jiajing supported a series of Daoist alchemists who promised to make him immortal, though none was successful), but Buddhism as an institution would not regain the authority it had enjoyed under the Yuan or early Ming. Daoism, far weaker as an institutional religion, would not even come close. When someone in 1403 presented the newly enthroned Yongle emperor with a set of Daoist texts the emperor shot back, "I use only the Five [Confucian] Classics to rule the realm. Of what use are Daoist classics?" Fourteen years later, a Daoist presented Yongle with an elixir of cinnabar. The emperor ordered the man to eat his own potion and commanded that all his books be destroyed.[11] The state did not ban Daoists, but it would not look to them for ideological support.

Buddhist monks were generally welcome in society, as they provided funeral and other religious services that were in demand. Officials understood this. "Buddhism and Daoism have been popular among the people since the Han and Tang dynasties and it would be difficult to do away with [them] completely," notes an official statement from the court. "The only thing to do is to be strict about maintaining the restrictions and agreements and not letting the two spread further."[12] An edict of 1418 limiting the number of monks to twenty per county would have further restricted Buddhism, had it been enforced. The fact that it wasn't suggests that, while the state may have worried about the economic loss caused by men fleeing into the clergy and thereby being released from corvée labor, the people were rather more worried about the strangling of religious life and the dwindling of monks qualified to conduct funerals.

Strict Confucians shared the anxiety expressed by officials at court. A scholar writing in north China in 1373 insisted that "Confucians do not talk about things related to the Buddha or the ancient Daoist patriarch Laozi. Using their propaganda about sin and fortune to transform ignorant customs is like using a torch to brighten the sunlight."[13] Confucian wisdom is the sun, Buddhist teachings a pale fire. As the Mongols had just been driven from north China five years prior to his making this comment, this aggressive posture may reflect his hope that Confucians would no longer be treated as just one among several religious technicians, as they had been in the Yuan, which had forced them to compete in the patronage market. He wanted Confucians, himself included, to be the leading ideologues and ritualists of the new order.

Support for Buddhist institutions surged among the gentry during the latter half of the sixteenth century, though this surge also provoked another wave of reaction on the part of conservative Confucians. Some objected to the free mingling of Buddhist and Confucian ideas, others to the donations that were going to Buddhist monasteries in preference to other needs. The losses of the Wanli Sloughs caused many to worry about the funneling of local largesse to Buddhism. "Today the realm has reached an extreme of poverty," observes a magistrate in north China writing in 1604. "If we wish to economize, nothing is better than cutting out extraneous expenses; among such expenses, nothing is more wasteful than constructing palatial buildings; and among palatial buildings, nothing is more wasteful than monasteries." This magistrate is painfully aware that sober Confucian principles lack the curb appeal of the fantasies about death and destruction that run through popular Buddhism.

> The delusions of this generation cannot be dispelled. The first delusion is not respecting parents at home but respecting spirits and Buddhas outside the home. The second is not trembling before state regulations but secretly fearing to go against the Buddhist dharma. The third is not mending what is right in front of your eyes but instead trying to mend what is off in the next life. The fourth is fighting over wealth with kinsmen while giving riches to priests and monks. Why don't even one or two ignorant men and women see this and return to orthodoxy?

In their private lives, many Confucian gentry accepted the presence of Buddhism in society and were even drawn into Buddhist religious practices. Some did this to placate their mothers or wives, but their interest was also stimulated by the efforts of eminent Buddhist monks who found their way into the intellectual world of the gentry and nurtured elite lay Buddhism. One of these was the Hangzhou tiger-queller Zhuhong, who gathered about him a glittering array of the brightest Jiangnan intellectuals of the Wanli era and instructed them in such pious Buddhist observances as releasing captured animals back into the wild, a practice known as "releasing life." Many turned to Buddhism to evade the fractious politics of the Wanli court.[14]

Buddhism was stronger among the southern gentry than the northern, who were fewer in number, weaker in identity, and less endowed to engage in financial patronage. One northerner writing in 1559, still early in the rise of lay Buddhism among the gentry, viewed the southern enthusiasm for Buddhism with dismay. "In recent times," he observes, "Confucians attired according to their status are all conferring their appreciation on the Two Masters [Buddha and Laozi], and are even giving instruction in [the doctrine of] annihilation in order to appear eminent. How distressing this all is!" Not everyone thought so. Another northern Confucian, writing two decades later, points out defensively that the ritual regulations of the Ming state "do not forbid local officials from supporting Buddhist or Daoist doctrines or praying for long life," which allows him to argue that there is no reason to restrain "the gentry from going into retreat in the temples that Buddhists and Daoists have built on great mountains and in deep valleys, or to stop there on their touring to catch the view and carve inscriptions." The pilgrim was also the tourist.

Cosmology

People of the Yuan and Ming imagined the universe as an arrangement based on the ancient idea that Heaven was round and Earth was square. This meant that Earth was flat and that Heaven curved over it like a dome. Another, more ancient cosmology imagined Heaven not as a dome but as an egg-like sphere in the middle of which was Earth. Though this view circulated only among a minority of intellectuals, it would prove helpful for adapting new knowledge coming from Europe, as we shall see.

From the model of a round Heaven and a square Earth flowed the cartographic representation of terrestrial knowledge. Throughout the imperial era, cartographers expressed this axiom by squaring land masses in their maps. The standard map of China down through the Yuan and Ming squeezed the bulbous shape of China into a square. The main distortion was in the southeast quadrant, where the coastal provinces of Zhejiang and Fujian, rather than curving from Shanghai in the northeast to Hainan Island in the southwest, were stretched outward to fill the empty maritime space off the coast. This rendering does not mean that this is how people "saw" the shape of the realm, only how they coded it. Such a map may not have been correct in our sense, but it met their expectations.[15]

The cartographer Zhu Siben (1273–1337) resisted this model. Working from Song antecedents, Zhu spent a decade constructing two large wall maps, one a national map of China, the other a "Chinese and foreigners map" *(huayi tu)* depicting the world that extended beyond China's borders. What drew Zhu away from the square-Earth model may have been his use of the grid method to transfer knowledge of smaller geographical units onto a larger map. Zhu's work influenced the best cartography of the Ming, including the first comprehensive atlas of China, Luo Hongxian's *Enlarged Terrestrial Atlas (Guang yutu)* of 1555. Luo's work did not entirely escape the square-Earth model, some of which lingers visually in his *General Map of the Terrestrial World* (Fig. 14). Within the traditions that made the realm recognizable, this was a coherent representation confirming fundamental conceptions of how people of the Ming looked at their world.

Conceptions, even the most fundamental, can shift at moments when conditions change. One of those moments came in the last half-century of the Ming with the arrival of Jesuit missionaries from Europe. European traders had been reaching China's shores since the second decade of the sixteenth century, but knowledge was transferred only at an individual level. This changed with the sustained presence of missionaries, who used their scientific knowledge to engage and intrigue Ming intellectuals. Cosmology posed the great challenge for both sides of this conversation. The Jesuits regarded Heaven as the outermost of nine concentric spheres surrounding the earth. As it happens, the image could be assimilated to the Chinese cosmic yoke model. Even the heavenly dome model could be made to correspond to this idea. The difficulty was what to do with the earth, which the Chinese saw as flat and the Jesuits as spherical.

The head of the China mission, Matteo Ricci (1552–1610), turned to maps to make the European argument for a spherical world. It is not intuitively obvious that planar images can be used to create a spherical form—which may explain why one of the first scientific books to be translated into China was Euclid's *Geometry.* It had not been intuitively obvious a century earlier in Europe either. Europeans had been experimenting with the cartographic problem of transposing curved surfaces onto flat planes since the time of Ptolomy, and they turned to the problem again in the fifteenth century as their mariners were traveling across the surface of the globe. Christopher Columbus did not discover that the earth was round. This he already understood. The question he could not answer without direct observation was how large the curve was, and it was larger than he thought. His initial error of believing Cuba to be Japan, and assuming that China lay just a bit further to the west, was the result of assuming that the globe had a smaller circumference than it actually had.

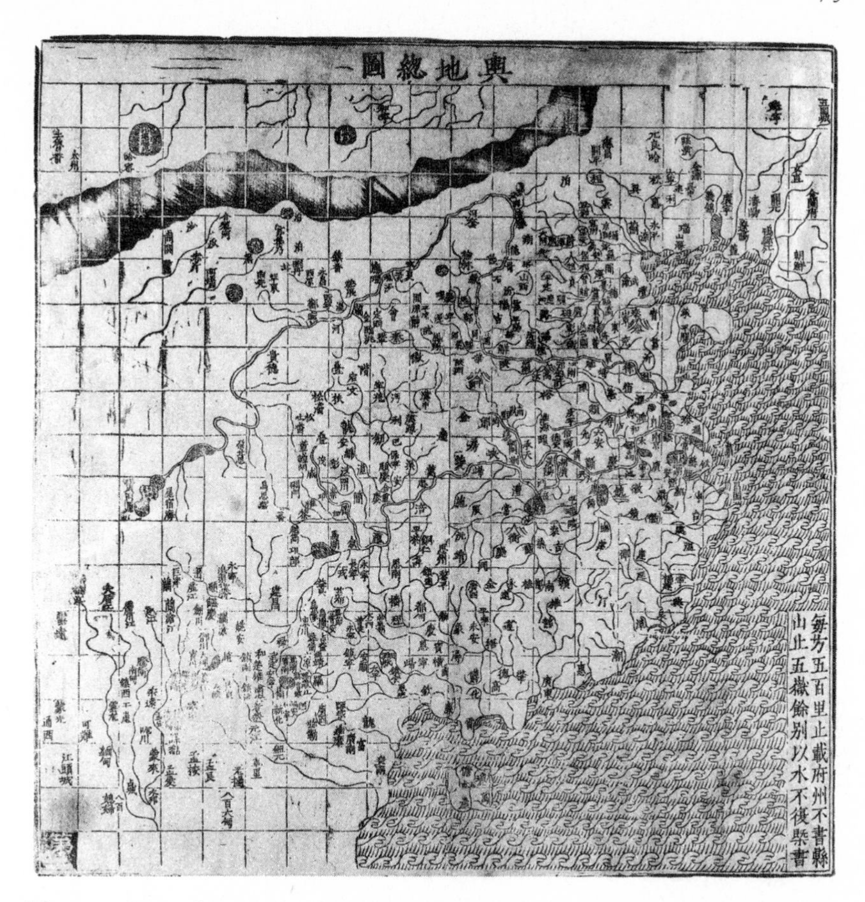

Fig. 14 *General Map of the Terrestrial World.* From the *Enlarged Terrestrial Atlas (Guang yutu)* of Luo Hongxian (1504–1564), 1555. The cartouche in the bottom right corner reads: "As each square measures 500 *li* [288 km], only prefectures and subprefectures are shown, not counties; of mountains, only the five marchmounts. Of other features, except for rivers, nothing further is marked." Note that the Yellow River flows into the Huai River south of the Shandong peninsula rather than following its current bed, which flows north of the peninsula. The black band across the upper left of the map marks the Gobi Desert. Of offshore islands, only Hainan Island in the far south is marked.

Soon after Columbus's voyages, European cartographers set to work realigning existing knowledge to his findings. The first map to name America, Martin Waldseemüller's 1507 map of the world, attempts a complex curvilinear perspective that Waldseemüller abandons in his later maps in favor of the greater visual clarity of rectilinearity, in which the lines of longitude and latitude meet uniformly at 90° to form a grid. It was a simplification that made visual sense, and is still widely used today. Between the rectilinear and the curvilinear emerged a compromise, the Mollweide or pseudocylindrical projection, which kept lines of latitude straight but allowed the lines of longitude to curve, and at an ever greater curvature the more they departed from the prime meridian. This projection suppresses curvilinearity at the center of the map but allows distortion to creep in toward the edges. Popularized by the atlases of Abraham Ortelius, this was the projection that the Jesuits took with them to China at the end of the sixteenth century.

Ricci produced as many as eight different world maps based on three different projections. It was the pseudocylindrical projection—a method that elongated the northern and southern zones of the globe to form a continuous band from east to west—that Chinese publishers reproduced most widely. As one of these publishers, Zhang Huang, tried to explain to his bewildered readers in his 1613 encyclopedia, *Illustrations and Texts*, "this map was originally designed for making a spherical globe, but it had to be cut to make a plane so that it could fit into a bound book."[16] The idea was hypothetical at the time, as no one had yet made a globe in China. That took another decade before two Jesuits produced the first known globe in Chinese.

Zhang Huang had no difficulty coming to terms with the new geographical knowledge and set himself the task of explaining it to others in his encyclopedia.[17] He offers an argument by logic. A lesson he takes from the Europeans, he notes, is the idea that the earth is finite. To set up his argument, he first quotes the Song philosopher Lu Jiuyuan (1139–1194), who debated with Zhu Xi about the nature of reality. Lu's view had been revived in the sixteenth century by followers of Wang Yangming, who championed the importance of relying on innate moral knowledge as much as on book learning to understand the world. One concept that had appealed to Lu was limitlessness, but Zhang declares it to be purely the effect of perspective. "If you look at the Earth by standing in China and gazing only as far as the four seas, it seems limitless. Even if you gaze as far as the Lesser Western [Indian] Ocean, it is still possible to imagine that the Earth is limitless," he allows, perhaps nodding to the limits of Lu's knowledge back in the Song dynasty. "But if you extend your gaze all the way to the Great Western [Atlantic] Ocean or the Far West [Europe], this is a distance that can be measured, so it cannot be called limitless. This map"—referring to Ricci's world map by pseudocylindrical projection—"demonstrates the measurability of the earth."

His conclusion: "The world is finite." And if Earth is finite, it must be a sphere. Recognizing that a popular encyclopedia is not the place for complex mathematical proofs, Zhang ends with the rhetorical question, "Is it only true because I believe it?" His answer is no. "That's just the way things are."[18]

Ricci's co-translator of Euclid and his most eminent disciple, Xu Guangqi (1562–1633), was an enthusiast of the European projections. He accepted that those who looked at Ricci's map for the first time would be puzzled, but he insisted to readers that the idea of Heaven and Earth both being spherical was "as easy as multiplying two times five and getting ten."[19] Xu uses geometrical calculations derived from Euclidean geometry to explain how it all works—explanations that must have baffled his readers. Rhetorically more effective are his citations from Song and pre-Song texts, which he selects and then construes as consistent with these calculations—though in most cases his interpretation is not really consistent with the original meaning of the text he cites.

For Li Zhizao (1565–1630), another high-status convert to Catholicism, it was not logic but experience that won the argument. Li met Ricci in Beijing in 1601, and on that first visit was shown the world map. Ricci explained to him that the map was the shape of a deformed cylinder rather than a rectangle because of the sphericity of the earth. Li was stunned, and very soon convinced. Prior to this map, "there was no good edition of a world map" in China, he wrote the following year. He notes that Ricci uses the same grid method championed by Zhu Siben and Luo Hongxian, thereby aligning him with the best traditions of Chinese science. He then goes on the attack. Compare it carefully with the map in the *Unification Gazetteer of the Great Ming* (an easy target, as it has to be one of the sloppiest national maps the Ming ever printed), he says, and you will see how much more accurate Ricci's is. The difference lies in his experience of the world he represents. "The man who composed this map did not rely on written records but personally visited these places. Written records are worth consulting only for changes over time; they cannot completely record everything topographical. Routes twist and turn without following the measuring string of the surveyor," after all. The natural inadequacy of maps is only that much worse in Chinese maps of the world beyond the Ming boundaries. Ricci has experienced these places, and therefore his version must be trusted over Chinese versions.

In a culture that revered the written word, the demotion of written records to the status of reference works was not likely to be received with equanimity. Even harder to accept was the proposal that a foreigner's version of the world should replace the Chinese version. But the trump was experience. "The Gentleman from the Far West," Ricci's title for himself, "has sailed the seas, personally crossed below the equator, directly viewed both pole stars, gone as far south as Great Wave Mountain [the Cape of Good Hope], and seen for himself that, from the perpendicular of the Southern Pole Star at its zenith, the earth is tilted 36°. Among the ancients who surveyed, is there one who has gone so far?"[20] The answer is no. Chinese ancients must bow to European moderns.

Li Zhizao, aware that he has not been able to supplant the square Earth axiom, tries a rhetorical gambit five years later. "The shape of the earth is indeed spherical," he declares, "but its virtue is square."[21] Another encyclopedist of the period, Wang Qi, picks up this argument in his *Illustrated Congress of the Three Powers* (referring to Heaven, Earth, and humankind). After reviving the cosmic yolk argument, he declares that "those who say the Earth is square are referring to its stable and immovable nature, not its physical shape." He then reminds his readers of the perfect correspondence between Heaven and Earth that shows both to be spheres. "Heaven already surrounds the Earth, which is how the two respond to each other. Heaven has south and north poles, and so does the earth; Heaven is divided into 360, and so is the Earth."[22]

The spherical Earth made inroads because of a factor internal to the Ming intellectual world: the willingness of Wanli-era intellectuals to *ge wu*, to investigate things. They were able to recognize that this is what the Jesuits, clearly well educated in mathematics and astronomy, had themselves done to reach their conclusions about the world. Using a method (spherical geometry) and an instrument (the telescope) superior to what Ming scholars had at their disposal, the Jesuits showed their Ming colleagues how they investigated Heaven and Earth. This gave their knowledge considerable credit, regardless of whether it disrupted the basic axiom of a round Heaven and a square Earth. The axiom was less important than the evidence that there existed a model that better accorded with observation. The Milky Way was thought to be a cloud formation below the moon, but the telescope revealed it to be a swath of thousands of stars beyond it.[23] With that observation, the belief that had kept the Milky Way below the moon had to be abandoned: there was simply no other choice. Within a decade, Galileo would use the same technology to contradict the basic axiom that the Jesuits were teaching Ming Chinese, that the Earth was at the center of a spherical universe. Yet again cosmologies at both ends of Eurasia had to change.

Moral Autonomy

Ming beliefs were not changing solely because a few Europeans initiated conversations with Wanli-era intellectuals, though their influence was significant. They were in flux because of pressures within the culture itself: political demoralization in the Wanli and Tianqi courts, rapid commercialization, status erosion, military emergencies on the borders, and environmental downturns. Under these conditions, some came to believe that the old certainties would not hold. They searched for new ways of understanding the world, and usually chose to do so outside the realm of official service. One such man was Li Zhi (1529–1602). Descended from a Muslim trading family in Quanzhou, Li had pursued the conventional course of passing the examinations and holding posts in the bureaucracy. In mid-life he retired from service and turned to philosophical speculation, which among other things led to a meeting with Matteo Ricci. In his later years Li took the tonsure and robes, though not the formal vows, of a Buddhist monk. He came to be seen as an iconoclast, enthusiastically taken up by the younger generation and spurned by the elder.

Li wrote a great deal, and by getting his writings into print made sure that his ideas circulated as widely as possible. Rather than survey everything he wrote, let us focus on one small set of writings, his correspondence with Geng Dingxiang (1524–1594). A considerable intellect in his own right, Geng was an early friend and later patron. But he was also a high official mindful of the political repercussions of radical philosophy, which did not concern Li. Their letters, which both men preserved in their published works, open at a late stage in the development of their disagreements. The correspondence traces both the unfolding of their ideas and the collapse of their friendship.[24]

The earliest surviving letter from Li to Geng, dated April 1584, disputes the relevance of Confucius to philosophy. For Li, the goal of study is not to understand Confucius; it is to understand the Way. Geng's "family teaching," on the other hand, promotes the method of working step by step through the classic teachings of the sages. Li disagrees with this method. "Confucius never instructed anyone to learn from Confucius," he insists, and quotes Confucius' advice that "the practice of benevolence arises from oneself."[25] He challenges what he sees as an unthinking acceptance of Confucius as the authority in all matters because he worries that the Confucian educational program has lost its way. Teachers and officials "use virtuous conduct and ritual to limit people's minds, and administrative methods and punishments to constrict their bodies." The task of education, and by extension the task of ruling, is not to force people to conform to philosophical or state regimens but to shape their natural impulses toward goodness. "Cold can prevent glue from sticking, but you cannot prevent people from rushing off to the towns; heat can melt metal, but you cannot melt the hearts of competitive people."

At this point, Li Zhi makes an unusual suggestion. Rather than curb selfish behavior with force or restraint, why not mobilize self-interest? "Allow everyone to go after what he likes, have everyone devote himself to what he is good at, and there will not be a single person who is without his function. How easy to rule in this way!" Toward the end of his letter, Li praises Geng for being earnest in his practice of moral cultivation, but gently asks him to accept that different views are possible. "Not everyone has to be like you," Li reminds Geng. "I respect you, but I don't have to imitate you."

In the earliest of the letters to survive on his side of the correspondence, Geng denies that the ancients had unalterable answers to everything. He sorts their models into two sets, those that have changed over time and those that have perdured for thousands of years. The former are based on experience and therefore mutable, whereas the latter were formulated following the rules established by Heaven. Times change, but underlying patterns do not. "The ancients expended much effort examining patterns so that you and I might have adequate shelter and sustenance," Geng writes. "They instructed us in ethics so that we might avoid being like wild animals." These instructions conform to what he calls "the obligation to action" or "the feeling of not being able to stop oneself" when moral action is demanded. Li's move toward Buddhism has led him to abandon "the mind's deep feeling of obligation to action."

Li's tone in the next letter hardens. "You follow old paths and tread in earlier footsteps," Li declares, whereas he, Li, is an impetuous soul who "like a phoenix flies at great heights" in order to "hear the Way." Better to be "impetuous and uncompromising" rather than "sanctimoniously orthodox," he says, quoting Confucius.[26] "There exist impetuous and uncompromising people who have not heard the Way, but never has there existed one who could hear the Way but was not impetuous and uncompromising." How could Geng hope to "hear the Way" if he continued to adhere so closely to proper behavior? Li Zhi is ready to abandon Confucianism, and Geng is appalled. In his reply, he complains that "the Zen fanatics of the present age" neglect study and simply pretend they have already entered a mystery beyond anything that mere Confucians could imagine. He blames Buddhism for giving people an easy alternative that produces no results. The serious practice of Buddhism may be difficult, but it is far inferior to the way of Confucius.

Li responds in turn by saying that Geng's "obligation to action" is like schooling younger brothers and sons to be pious and deferential, whereas his illuminates the minds of adults. Geng's advice is like rain that soaks everything without being sought; whereas his is like snow that gradually releases moisture to ease a drought. Geng is a village schoolmaster drilling his pupils without getting any results; he is a general who dispatches a squad to capture the other side's king, using little effort to great effect.

The correspondence spirals out of control. Li accuses Geng of having failed to intervene to save a mutual friend, the highly independent philosopher He Xinyin (1517–1579), from execution ten years earlier. At the time Geng had the ear of Senior Grand Secretary Zhang Juzheng—but did nothing to intervene. Stung, Geng writes back and asks what Li thought he was doing twenty-five years earlier when he went home to Fujian to observe mourning for his grandfather and let his two youngest daughters die of hunger back in Beijing. Philosophical difference shifts to personal recrimination and their relationship breaks down.

Li believed passionately in the capacity of the individual to find his own way to the truth; Geng believed firmly in the value of following the wisdom of the ancient Classics in order to apprehend the truth. The difference pitted them against each other to a degree neither could have imagined at an earlier stage in their friendship. In the long run, it would be Geng's position that most Neo-Confucians ended up supporting, and Li's they would reject. Li's views seemed only to lead away from the moral core that supported the obligations of statecraft, not to open a path toward a stable moral future. The two men came to a cautious reconciliation in Geng's last years, but it was not easy.

A decade after their correspondence ended, Li Zhi got caught in a political crossfire around another chief grand secretary who was struggling to parry an attack from critics. Targeted as a protégé of one of his opponents, Li ended up in a Beijing prison facing impeachment for the moral impropriety of having accepted a female student. Even though it was a trumped-up charge arising from a factional struggle inside the court, everyone took it seriously. A friend attempted to defend him by downplaying his ideas, saying, "Everyone has his own view, so how can they be all the same?" Uniformity, he asserted, is intellectually unhealthy: "If all philosophers who agreed completely were judged acceptable, and all who disagreed unacceptable, then Zhu Xi and Lu Jiuyuan would not have had their debate."[27]

We cannot say whether this plea was heard or not. In any case, the charge against Li was dismissed. He was ordered to be released into the custody of a friend. Before the order reached the prison, however, Li slipped into despair and committed suicide. Modern historians regard him as a martyr to intellectual autonomy; to his contemporaries, he was a crazy old man who ended his life as controversially as he had lived it.

Sameness and Difference

Difference is a standard means of organizing our understanding of the variety of the world. As the popular Cantonese adage put it, "Go a thousand *li* and customs are dissimilar, go a hundred *li* and habits are different."[28] This was as true within China as beyond the imperial borders. The Ming state aspired differently. "The highest realization of ruling is that the country should be without customary differences," opines a writer in eastern Guangdong in 1519. "Thanks today to the emperor's spiritual power, all within the four seas is unified as one, so how could there be anything that could be called differences in customs?"[29] According to this vision, natural difference was no ground for doubting the unity of the nation and of political or cultural identity within it. When the Ming founder complained of the southwest that "customs and practices differ," it was not to admire local variation, but to put his officials on notice that something should be done to replace this difference with uniformity.[30]

Nonetheless, an awareness of difference was ingrained in the Ming mind as a fundamental experiential fact. Troubled by the evident differences that denied the underlying sameness of the world, intellectuals of the Wanli generation struggled to assert unity. As Peter Bol has observed in his history of Neo-Confucianism, "The belief in unity—of the cosmos as an organic system, of antiquity and an integrated social order, of doctrine as universal and unchanging, and of the mind's experience of oneness—was at odds with the world in which Neo-Confucians lived."[31] It was a world in which opinions even among educated men differed wildly, Buddhism and Daoism were as vigorous as ever, and the realm was surrounded by people who were not the same in habits or beliefs. The commitment to the ideal of unity faced the uphill gradient of the real world, and was espoused no less firmly because of that.

The notion that not everyone had to think the same way was not a position Wanli intellectuals found easy to defend. Li had tried to defend it with Geng Dingxiang by invoking the logic of the unity of the Three Teachings: one could vary one's approach to the truth without compromising the truth itself. A preference for one path did not negate other paths. Orthodox Confucians, however, were horrified to see the teachings of Confucius equated with the teachings of the Buddha, and this view gained strength as things fell apart toward the end of the dynasty. The father of the great philosopher Wang Fuzhi (1619–1692) was one of these end-of-dynasty Confucians who deplored what he saw as the erosion of Confucian authority. As his son later wrote, "My father to the end of his life never once offered a single bow of respect before a statue of the Buddha or of Lao Zi." Wang put his father's attitude on record to forestall any misunderstanding of his decision to work with Buddhist monks in 1644 to clear the countryside of corpses after the army of rebel leader Zhang Xianzhong pillaged its way through southern Huguang.[32] For the strict Confucian, Buddhists were not problematic or offensive; they simply belonged to a different category. Men like this in the generation who lived through the fall of the Ming looked back to the blurring of boundaries in the age of Li Zhi as the moment that tipped them into the abyss of dynastic collapse. The more radical of Wang Yangming's followers had gone so far as to declare that everyone was a sage: that everyone at base was the same. Nothing could be further from the Confucian project, which was to produce an exact structure of moral and social distinctions. But it was a proposal that some bright minds in the tumultuous cauldron of ideas in the Wanli era were willing to entertain.

Some of these bright minds also ended up in conversation with the Jesuits, which led them even further in the defense of universality as a higher principle than difference. Lu Jiuyuan had a saying they liked to repeat. "Eastern sea, western sea: mind is the same, truth is the same." Li Zhizao, a convert, quotes it to plead that though Chinese and European theories of geography and cosmology differ, their methods and findings are not incommensurable.[33] Both cultures have learned to measure the surface of the earth and observe the sky in order to analyze the composition of the universe. Since the universe is the same, their findings must ultimately be the same, and their methods compatible.

The Catholic convert Xu Guangqi makes this point in an essay defending Christianity, "An Outline of the True Way." Both Chinese and non-Chinese are subject to common conditions, or as he phrases it, "The same wind blows across the four seas." When one recognizes that all existence has a common source, "what other is there, and what self?"[34] Xu was in part defending himself against the charge that Christianity was fundamentally incompatible with Chinese culture, and that meant refusing to accept one culture's truths over another's. In his terms, "there is nothing that is not the same." Christianity did not prompt him to his idea so much as lead him to realize the need to create space for a different set of beliefs. The basic idea already existed within the Confucian tradition going back to Lu Jiuyuan. Lu had sought to counter Neo-Confucian hyperrationalism by investing the subjective experience of the individual with greater moral authority. It was a view to which Wang Yangming in the mid-Ming had been sympathetic, and to which Wanli intellectuals such as Li Zhi subscribed.

Those hostile to influences from the outside preferred to raise the specter of difference. Chinese who absorbed European knowledge, especially the knowledge of Heaven that the Jesuits regarded as central to Christianity, were accused of honoring a "different principle" and following a "different teaching," falling just short of the state crime of heterodoxy. In hypothetical dialogues he wrote to dispel doubts about Christianity, another eminent convert, Yang Tingyun, tackles the objection of difference. His fictional conversation opens with his interlocutor asking, "Are the writings coming from the West the same or different compared to those of our China?" "Largely the same" is his answer. If that is so, the interlocutor presses, why embrace Christianity when the Chinese tradition already includes everything that is in the European? Isn't European knowledge simply unnecessary?

"Not so," Yang replies. "What is meant by 'sameness' refers to the unity of that which controls everything and which is the ancestor of everything." Yang then goes back to fill out his remark that Chinese and Western traditions are "largely the same" by providing a checklist of four points of potential philosophical sameness and difference, or as he rephrases it in the checklist, "not-yet sameness." These are worth investigating—or in his language, "cutting up," the same term Zhang Huang uses to describe the cartographer's work of cutting up a sphere and laying it on a plane—in order to understand what Europe has to offer. This is the point of Yang's exercise: not to measure out strengths and weaknesses on each side but to show where China might benefit from the learning coming from Europe. Just because this knowledge was unavailable to the ancients does not mean it should be rejected. Yang goes on to insist that "the differences really do not harm the similarities, and the similarities do not diminish the differences." He offers an instructive local example: "It is the same as in an examination hall where the students have the same theme but their essays distinguish themselves naturally in their degrees of elegance and ingenuity. So why," he asks, "should the theme be considered different?"[35]

Sameness was not just a philosophical idea. It was how some people at this moment of cultural flux chose to respond to the challenge of identity when another philosophy confronted them. The Yuan had governed by supporting the notion of a multi-ethnic state. Not only did it fail to unify the people, but it installed an ethnic order that insisted on a fundamentalism of difference. The Ming abandoned that multi-ethnic approach, choosing to unify only those who were already a unity, the Hua or "Chinese." The Yuan had claimed national sameness but practiced ethnic difference. The Ming practiced both ethnic and national sameness. As time wore on, though, the toxic residue of sameness in the popular mind was xenophobia: a fear of those who are not the same. Refusing to grant sameness to Europeans, or any other outsiders, made it difficult for most officials in the Ming court to develop an informed accommodation with the outside world. The Manchu victory in 1644 seemed to confirm that the xenophobes had been right. Those who sought to unify all traditions and beliefs lost the moral ground on which to defend their vision of a more inclusive world.

8

THE BUSINESS OF THINGS

THE YUAN and Ming were awash in things. From the vast storehouses of precious objects and artworks locked away in the imperial palace, to the elegantly furnished residences of the well-to-do, to the one- and two-room homes of ordinary farmers, people accumulated around them the objects they needed, or believed they needed, to lead their lives and conduct their affairs. These things were as simple as a chopstick or as common as a teapot, as finely crafted as an eggshell-thin teacup from the Chenghua era (1465–1487) or as elaborate as a sheet of jade carved with minuscule figures playing on a cloud-strewn landscape. Some of these things were abundant and others extremely rare, some breathtakingly expensive and others ridiculously cheap. They defined the material world in which the people of the Yuan and Ming lived their lives. It was a world in which commodities were produced, circulated, and consumed in a variety and on a scale that no culture had yet experienced. A poor farmer might be able to afford one major purchase a year, while a prince of the Zhu family might have warehouses of goods exceeding any notion of need. Things fit themselves into the spaces that buyers could afford, and did much to define their worlds.

Household Possessions

Records of who owned what in the Yuan dynasty are sparse. Marco Polo supplies us with glimpses of luxuries he saw in Khubilai's palace, such as the enormous wine chest in the main banqueting hall, "each side being three paces in length, elaborately carved with figures of animals finely wrought in gold." At banquets, he tells us, the khan's servants produced "such a store of vessels of gold and silver that no one who did not see it with his own eyes could well believe it." Impressive, but that is the whole point. Polo is not really trying to give us a complete account of the things with which a powerful ruler surrounded himself. His purpose is to astound us with the scale and value that these things represent, not to furnish later historians with an inventory of what was actually there. The only inventories of possessions we have from the Yuan dynasty are of art collections in the hands of forty-seven private collectors living in Hangzhou during Khubilai's reign, to which we will return in due course. This is not the sort of record from which it is possible to build up an image of the things that filled Yuan households.

We are better served for the Ming. The most famous list of Ming possessions is the inventory of what was confiscated from the household of Yan Song (1480–1565). Yan was a grand secretary whose monopoly of power during the last two decades of the Jiajing emperor's reign earned him the envy and dislike of the entire capital bureaucracy—and earned him as well a great trove of gifts and purchases that the state duly confiscated before he died in disgrace. The charges laid against the grand secretary, other than being sycophantic to the emperor, were probably unfair exaggerations, though the charges against his son Yan Shifan (1513–1565) for using his father's position to enrich himself and run roughshod over the people of their home county were deserved. The inventory that was drawn up at the time of confiscation in 1562 is preserved. It is as much a political document as an objective list of possessions, but it does provide us with a glimpse of life in one of the wealthiest households of the country.

The Yan Song confiscation shows the upper end of what a wealthy consumer could hope to own: vessels fashioned from gold, silver, and jade; antiques, mostly bronze; objects manufactured from rare materials such as coral, rhinoceros horn, and ivory; fancy belts (these, like men's ties today, were the significant fashion accessory for men wearing robes); fine cloth, especially silk, and the clothes tailored from it; musical instruments, some of them antique; ink stones and writing accessories; folding screens; wooden beds inlaid with precious materials; calligraphy, rubbings of stone inscriptions, and paintings of all sorts; and finally, a great library of books. And these are only the objects that the state confiscated directly. The Yan family was also deprived of many more things that were put up for forced sale and entered into a different inventory.

This second list included objects of a more generic sort: vessels and utensils, textiles and clothing, furniture and bedding, musical instruments and books—all of them of fine quality and high price, without doubt, but not antiques or important masterpieces or cultural treasures. These were the things that a rich family liked to have to use on a daily basis. As noted, the confiscation was a political act. This two-part inventory existed to prove that Yan had been morally unfit to serve as a grand secretary. Still, the art historian Craig Clunas doubts that Yan's political enemies doctored the document. "The bland bureaucratic language in which it was written is not that of prurient excitement but rather of a dispassionate listing of seemingly inexhaustible riches." These are the actual things that a real family, albeit an atypical one, owned.[1]

The next set of inventories we have to consider are those that survive in contracts drawn up to divide family possessions when a household divided, usually on the death of the household head. Five of these have survived from Huizhou, the home prefecture of many of the greatest merchant families of the Ming dynasty in the hills south of Nanjing.[2] These families were not in Yan Song's league, but they were sufficiently well off to have possessions enough to dispose of through contract. The items in the possession of the Wu family when it divided in 1475 suggests a household that was only mildly prosperous. The contract lists a carpet, 2 sitting mats, a decorated lantern with stand, a pair of old bronze flower vases, 4 lacquer trays, an abacus, a painting, a chest, a clothes rack, and a wine chest. The family also owned a grinding trough, a level, a saw, 3 sedan chairs, and what appears to be some sort of petard or firearm. The Wus were far from poor—owning 3 sedan chairs to go about in is a clear sign of that—but their wealth was such as placed them well below the upper elite.

The Yu family, which divided in 1634, has left an inventory that suggests what an ordinary family might have owned a century and a half later. Their inventory includes 10 tables of various shapes and sizes, 2 beds, an incense stand, 12 stools, 12 chairs, 3 sets of steps, plus an old stand for a transverse string instrument known as a *qin*. Many of these are listed on the inventory as "old." So the Yu family owned more objects than the Wus, but they were roughly at the same level of wealth compared to their contemporaries. Huizhou in 1634 was a far more prosperous place than it had been in 1475, and expectations about appropriate furnishing would have changed. That a modest family should now own a stand for a qin, the trademark instrument of the gentleman, indicates that cultural expectations had risen as well.

The objects itemized in the contract of division that the Sun family drew up in 1612 shows what a much wealthier family could possess. The Suns were merchants, and their trade had brought them to a prosperous state. All three sons having married, the Suns decided to divide their joint possessions, which were considerable: gold and silver vessels, objects in bronze and tin, paintings, porcelain, and no fewer than 180 pieces of furniture. We shall examine the furniture more closely later. For now, simply note the number. This was a lot of furniture, and indicates how many objects a wealthy merchant family could expect to have around its residence.

The Chengs, another Huizhou family, earned a fortune operating a chain of eight pawnshops up the Yangzi River valley, and they possessed wealth beyond the Suns. Their inventory of household furnishings drawn up in 1629 lists only 53 items, but it is a curious list: 15 incense tables, 34 lacquer boxes, 3 screens, and a wine chest with bronze inlay. If the Chengs at first glance appear to own many fewer things than the Suns, it is simply that their division contract lists only the really expensive objects. Ordinary furniture and daily-use items were not included. In a family like this, they were of no account.

The inventory that helps most in sketching a complete picture of what a large household might have owned is the most unusual one, and again the result of a confiscation: the Jesuit compound in Nanjing. The confiscation occurred in 1617, as part of an inquiry into Jesuit activities in the southern capital. The two Jesuit missionaries in Nanjing, Alfonso Vagnone and Álvaro de Semedo, were arrested in 1616 along with seventeen Chinese associates, most of whom lived with them in their house. The charge was sedition. The following spring, as part of the investigation, their house was searched and sealed, at which time a thorough inventory of its contents was drawn up. Matteo Ricci had bought the house in 1599, which meant seventeen years of habitation, and seventeen years' accumulation of things. Some of the contents—the organ, for instance, and the clock (not working) in its wooden case—were peculiar to the Europeans who lived there. But most of the inventoried objects were the sorts of things any large household would own.

Three separate lists were drawn up: one of 67 foreign objects that were returned to the Jesuits to remove from the country when they were expelled; a second list of 1,330 items of furniture and other household furnishings, which were of no interest to the state and were therefore sold off; and a third inventory of 1,370 items confiscated in connection with the sedition charge. These were mostly books (850 volumes), but they also included printing blocks, documents, maps, astronomical instruments, crucifixes (which authorities regarded as akin to voodoo dolls), and objects decorated with the restricted motif of dragons.[3] This was a lot of stuff. Despite the comment of one European visitor that the Nanjing house was not particularly nice, the lists suggest that it was furnished perhaps not luxuriously but at least well. It is hardly surprising that it should have been. The Jesuits had to accommodate at least a dozen residents at any one time, and they had to entertain callers in style.

It is hard to know where to begin with so many things. We can start with the furniture: 40 tables, 61 chairs, 34 benches and stools, 5 bookcases and 11 bookshelves (plus 2 loose shelves), 13 cupboards, 9 camp beds, 3 regular beds, 2 canopy beds, and any number of cases and cabinets. The porcelain amounted to 326 pieces, plus two large porcelain censers. There were bolts of cloth, clothes and handkerchiefs, curtains and coverlets, kitchen utensils and trays, cabinets and storage boxes, all in great number. The copper items alone included a basin for boiling water, a vessel for keeping tea warm, a heated sink, 7 trays, 4 censers (2 with copper stands), 2 cooking pots, 2 panels, and 2 spearheads. The pewterware consisted of a wine kettle, 6 wine carafes, 4 teapots, 3 jars, a lampstand, and 10 candlesticks. Among the larger objects there were 3 sedan chairs plus curtains and screens and a pair of poles to carry them, 3 mule litters, 3 iron stoves and 1 iron heater, plus 2 carpets, one wool and one hemp. The collection of tools included 4 saws, 2 steelyards, and a lathe. In addition to these non-perishables, the house was stocked with 400 liters of rice (enough to feed a dozen people for three months), a bucket of salted eggs, 10 loads of firewood, and 10 jugs of liquor.

The inventory was not treated as a reproach to the Jesuits. There was nothing here to indict the missionaries as extravagant or corrupt. The astronomical instruments were suspect, not because they were extravagant but only because scanning the skies was the prerogative of court astronomers. So by the standards of the time, this was simply a well-stocked house, perhaps not all that different in the number and quality of its furnishings from any large and reasonably prosperous household in Nanjing in the decade of the 1610s—a prosperous decade to which we will shortly return.

Connoisseurship

Things are not just inert objects that simply do what we make them do. They carry meanings, meanings sometimes so powerful that they overwhelm their use entirely. The double life of things is most easily spotted at the upper end of consumption. The court, for instance, had to be filled with objects of the finest quality, not because an expensive elegant stool was more useful than a cheap sturdy one, but because it was doing more work than supporting the rear end of the person sitting on it. It was what it was because it had to publicize the wealth and elegance that the court was expected to embody.

The Yuan and Ming courts were accordingly major consumers of luxury objects: paintings to be hung on walls, furniture to be sat on, place settings ordered from the porcelain kilns in Jingdezhen, silks to dress themselves and their families, elegantly bound books to read and to present to loyal subordinates. The scale of courtly consumption was vast. An entire apparatus of state workshops, some of them within the precincts of the palace itself, some in key manufacturing cities such as Suzhou and Hangzhou, came into being to manufacture the luxury objects the court commanded. Popular taste followed suit, of course. People outside the imperial family eyed these luxuries for themselves and connived to consume them, though they could only do so within some very particular rules—such as making sure that whenever you bought something with a dragon on it, that dragon's feet sported only four claws instead of five (Fig. 15). Recall that the discovery of bowls decorated with dragons counted against the Jesuits.

Taste was not a one-way conveyor belt extending from the court to society. Some people might wish to imitate the emperor by acquiring the objects he consumed, or more likely knock-offs of the real things, but to men of discrimination, this was a losing game. Better to set your own standards—and this is what the gentry did, developing styles that accorded with their own consumption preferences. These hinged not on what was costly and conspicuous (though it was always nice to be noticed, especially when you had paid a lot for the thing being consumed) but on what was elegant. Elegance was a tough criterion to master, tough enough to stump the *nouveaux riches*. It could even be tough enough to put emperors at a disadvantage, which was the point. What did an emperor have except Heaven's mandate, a security apparatus, and an apparently endless supply of cash? Without his Confucian tutors, he could

Fig. 15 A porcelain jar manufactured at Jingdezhen, its base marked with the date of the Jiajing era (1522–1566). The dragon's two visible feet show five claws, signifying that this object was produced for the imperial household, though it may well have ended up in the possession of others. Arthur M. Sackler Museum, Harvard University.

have no knowledge of the subjects of which men of good taste should be in full command: antiquities, painting, calligraphy, books, even comportment. Khubilai and Zhu Yuanzhang did not trouble themselves about mastering such arcana. Their descendants, many of whom came to the throne as children, did no better. They had tutors in such matters, but they listened to them half-heartedly. Compared to their Song predecessors, the thirty Yuan and Ming emperors stand out for their utter lack of cultural attainments. The exception is the Xuande emperor (r. 1426–1435), grandson of Yongle and great-grandson of the founder—the rare case of an emperor sufficiently absorbed in the culture of elegance to achieve real skill as a painter. But he is the only one.

In an economy based as much on taste as on money, the emperor as wealthy consumer had to cede place to the gentry connoisseur as elegant consumer. Emperors merely possessed things, whereas connoisseurs used them to express the highest ideas of their culture: thoughtful contemplation, aesthetic discernment, and good taste. The two practices of consumption—the conspicuous and the elegant—influenced each other, but largely went on in separate social realms. Thus, while Zhu Yuanzhang was furnishing his palace in Nanjing, a wealthy collector by the name of Cao Zhao was in the same city compiling a guide to collecting elegant artifacts, *Essential Criteria for Discriminating Antiquities* (*Gewu yaolun*), which taught gentry readers how to identify objects worth collecting and to appreciate them without being tainted by the urge to possess them.[4] The emperor would not have been interested. Still, whether merely acquisitive or deeply cultural, consumption had the powerful effect of stimulating the creation of an extraordinary oeuvre of art and artifacts that defines what most people think of as "Ming."

Connoisseurship did not begin in the Yuan dynasty, but it was greatly stimulated by the flood of objects released onto the market when the Yuan dispossessed the Song imperial and aristocratic families. Those anxious about preserving the heritage of the past—which the Mongol occupation threatened to annihilate—turned to the social practice of connoisseurship to remember the Song. Connoisseurship in fact emerged as an activity through which southerners still loyal to the Song and northerners serving the Mongols could meet and find common ground.[5] Connoisseurs of the early Yuan focused their attention on calligraphy and painting, which they revered as living links to their cultural past, though antiques and ceramics also crept into their collections. In the early Ming, the field of elegant consumption was limited by what was available. This changed in the sixteenth century as the commercial economy responded to demand. No longer content with the minimum necessary to operate a household, wealthier families began to amass possessions through which they could display their wealth and discernment, from ordinary luxuries, such as well-made furniture, porcelain dishes, and nicely printed books, to cultural luxuries protected by rarity and price: Shang bronzes, Tang paintings, Song imprints, early Ming porcelains, and calligraphy by masters ancient and modern.

Possessing such costly objects required wealth, education, and connections, and over the course of the Ming the number of buyers who met these requirements grew. As this happened, competition intensified between those who regarded themselves as the guardians of the best cultural traditions and those who were trying to force their way into polite society. The newly wealthy challenged restrictions that had once kept commoners in their place. Moral conservatives responded to this challenge by looking back and pinpointing a time when standards started to slip and the social order to decay. Conservatives in Jiangnan in the 1530s looked back to the 1460s as the time when greater prosperity seduced more people into valuing luxury consumption over ritual propriety.[6] Conservatives in Shandong and Fujian in the 1540s blamed the opening decade of the 1500s. In Henan and Zhejiang in the 1550s, they blamed the 1510s, the infamous Zhengde era. And that was where the recriminations stuck, preserving the Zhengde era—a busy era for dragons—as the universal scapegoat for all that had gone wrong in society.[7]

So much more was riding on the consumption of luxuries than the simple distinction between who could afford them and who could not. To know what a ritual wine goblet from the ancient Shang dynasty should look like had been an item of knowledge reserved for highly cultured gentry families. To be able to recognize the calligraphy of Mi Fu (1051–1107), considered the greatest calligrapher of all time, was the difference between entry into and exclusion from the world of the elite. But connoisseurship in a rigid status-based society entails more than just knowledge. It is a social activity for which people of like standing come together to appreciate objects of high value, and in so doing recognize and appreciate one another (Fig. 16).

Some of what the gentry collected was the work of men of their own class, but most was the work of artisans. The very best objects were created deep in the past, though by the later decades of the sixteenth century, contemporary artisans were achieving national reputations, with brand names that helped buyers navigate their way through the market. The emergence of branding could not have come about under the Yuan, when artisans were bound into government service. As the Ming gradually commuted service into cash payments, enterprising artisans escaped their bonded status and set themselves up as independent producers. Finding strength in numbers, artisans tended to cluster in the same part of town and eventually organized themselves into craft guilds to protect and regulate their collective interests.[8] The guildhalls they built usually took the guise of a temple dedicated to the craft's patron deity. In Suzhou, for example, metalworkers worshipped a craft ancestor they called the Old Master at the Old Master's Hall, and embroiderers' honored a Jiajing of-

Fig 16 *Enjoying Antiquities* by Du Jin, who painted in Nanjing during the Chenghua and Hongzhi reigns (1465–1505). Two collectors pore over a table of antiques, attended by four servants. The woman in the upper right is wrapping a *qin,* the gentleman's musical instrument. National Palace Museum, Taipei.

ficial who taught the women of his household to embroider by gathering at the Embroidery Pattern Master's Lodge.[9]

Despite the rise of skilled manufacturing, collectors believed that there were some objects, notably painting and calligraphy, that artisans could not make as well as amateurs could—themselves, in fact. Calligraphy was the art form that most directly expressed the spirit of its producer. A mere technician could never produce calligraphy of real quality. It had to come from the hands of the elite. Painting was a close second. Urban markets were full of paintings and calligraphy produced by artisan workshops, but the serious collector avoided such soulless items. They did allow that there were professional painters of talent at court, but they held the inspired amateur in higher estimation.[10] By insisting on this distinction, they elevated themselves as artists, not artisans.

To get a sense of the process by which people of the late Ming acquired the things that were meaningful to them, I suggest we turn away from inventories to running tallies, specifically to one running tally in particular kept by a gentleman by the name of Li Rihua (1565–1635). Li was born in the city of Jiaxing, a prefectural capital on the Yangzi delta between the commercial hub of Shanghai to the northeast, the cultural and commercial center of Suzhou to the northwest, and the former imperial capital of Hangzhou to the southwest. Li's family was not initially prosperous, though his orphaned father was able to build up property holdings sufficient to pay for his son's education and civil service examinations. Li passed his *juren* or Elevated Person degree in 1591 and his *jinshi* or Presented Scholar degree in 1592, and then won a strong first appointment as a prefectural judge. But the death of his mother in 1604 obliged him to leave office after a dozen years and go into mourning. The posture clearly suited him, for after the required twenty-seven months of mourning, Li used the excuse of having to care for his aging father to stay out of the increasingly fractious political realm for another two decades. He stayed home, kept his head down, and enjoyed the elite pastimes of his generation: painting, versifying, traveling, joining gatherings with like-minded friends, and engaging in the local politics.[11]

We know about Li Rihua because he kept a diary during his retirement, and eight of those years (1609–1616) miraculously survive. The work's title, *Diary from the Water-Tasting Studio* (*Weishui xuan riji*), reflects his passion for tea. Only the true connoisseur was the ability to distinguish good water from bad; anyone, after all, can taste the tea itself. Li could write, he could paint, and he had a fine calligraphic hand, but none of what he produced stands out from what hundreds of his contemporaries were producing. Only his diary sets him apart, for it shows us a gentleman of wealth engaging day by day in all the pursuits of his class. One of these was the pleasure of acquiring valued objects.

The 1610s were a decade of prosperity. Li could afford to collect the things that he deemed as embodying the best of his cultural tradition. Elegance was his first criterion. Distinguishing between the elegant and the vulgar was the most important act of connoisseurship. Second to that, and in a sense required by it, was distinguishing between the genuine and the fake. It was unthinkable that something elegant could be fake. By the same token, no matter how shoddy a piece might be, if it was the work of a prized artist or artisan, Li was quick to grant it elegance. When the attribution was weak, he was readier to detect its lack.

Li's greatest challenge in building up his collection of valued cultural objects was limited supply. There were simply not enough quality items on the market at any one time, which is a common condition of the luxury market. Friends and acquaintances might have items in their collections they were willing to part with, but these personal networks were insufficient for a collector of Li's ambition. He needed to work through commercial networks as well, and did so. Rarely did a week pass without Li noting in his diary that at least one dealer from any one of half a dozen major cities on the Yangzi delta stopped by with luxury objects to sell. His most constant supplier, however, was a local merchant we know only as Dealer Xia. Xia appears in the pages of Li's diary forty-two times over seven years, bringing Li goods that ranged from masterpieces to trash, hoping to make a sale just as much as Li was hoping to find a hidden treasure. Let them take us through the high end of the world of things toward the end of the Ming dynasty. We will focus on four things in particular: books, furniture, porcelain, and painting.

Books

As we have seen from the Yan Song and Jesuit confiscations, a wealthy household expected to own books in large numbers. As it happens, books were the one luxury commodity in which Dealer Xia did not deal. Books could end up in estate sales and job lots and circulate alongside other luxuries, but mainly specialized dealers handled them. Books, however, were not one thing but a vast array of things, from pulp fiction (which became hugely popular toward the end of the sixteenth century) to elegantly engraved editions of the Classics targeted to the high end of the market. When so many young men struggled through the examination system, as in the Ming, books were particularly prized as tools of advancement as well as objects of cultural reverence. Even so, the bulk of the trade aimed at a lower-brow readership.

The technology of book production was not difficult to master but required coordinating teams of artisans who specialized in each step of production. Once a manuscript had been finalized, scribes transcribed the text in mirror image onto blocks of pear wood, two consecutive pages to a block. Engravers then cut the blocks by carving away the surface of the wood around the characters so that the text stood higher than the background. Producing a 200-page book could involve the labor of 2 calligraphers, 3 copyists, and 6 engravers.[12] Printers then inked the woodblocks, spread sheets of folio paper over them, and pressed them onto the blocks to produce an imprint showing the two pages side by side. Binders folded these sheets in half so that the two "pages" on the block faced away from each other to make a "leaf," to use the Chinese word for what resulted. These were then stitched into paperback volumes, called fascicles. Cover-makers cut cardboard rectangles to the size of the folded leaf and the depth of the fascicles, then glued these into cloth slipcovers that enfolded several fascicles together. The result was a book.[13] A single title could be one fascicle, one cover, or many covers.

Scholars concerned to ensure that certain works were in circulation sometimes got into publishing. Over the Yuan and Ming their output declined relative to commercial publishing, which moved into ascendancy around the turn of the sixteenth century as literacy spread and readers grew in numbers. These developments reinforced each other, building an enormous market for books and buyers by the Wanli era. Unlike scholarly publishers, commercial publishers were concerned to find a readership (and buyership), not necessarily to express the cultural values that scholars sought to embody in their writing; and they were in business to make money. But commerce and scholarship could coincide. Some commercial publishers took on scholarly projects that they hoped would earn them a profit, and some scholars responded to market demand by writing and marketing popular works. That said, much of what went into print in the Wanli era and beyond was the work of professional hacks who cranked out stories, satires, potted histories, examination cribs, erotica, encyclopedias, and handbooks on everything from how to write a letter to how to administer a county.

We get a sense of the presence of books in gentry society from a sketch that the brilliant scholar Gu Yanwu (1613–1682) wrote of his own family's book-collecting practices. The family was from outside Shanghai. Gu begins by explaining that his great-great-grandfather started collecting books about 1520. This was still a time when the only imprints available were published in the princely palaces of the founder's heirs sequestered around the county, or by government offices, or by commercial publishers in Jianning prefecture in the Fujian interior. The kinds of books one could find in print were limited to the Classics, the standard histories, and the writings of the Neo-Confucians, which made for a fairly orthodox, even dull reading list. Gu's great-great-grandfather was nonetheless able to amass a collection of six to seven thousand volumes.

This library was destroyed during an attack by Japanese pirates in the 1550s. The man's son rebuilt the collection in the Wanli era when, Gu notes, "books were easier to acquire than formerly." By this time, the severest enemy of collecting was time itself, for Gu adds that the new purchases "did not include a single imprint from the early Ming or earlier."

Old books were becoming rarer and hence more expensive. As this happened, the rare book market took off. Buyers in that market were not just looking for a text; they were looking for a text that only they could possess, the rarer the better. Not having an unlimited book budget, Gu's great-grandfather declared a virtue of necessity, disdaining those who bought rare books just for the sake of owning expensive objects. "The books I have collected are for the texts they contain," he declared. "Skillful ivory ornaments and silk bindings are not what I care for." The collection was split four ways among his sons after he died. Gu's grandfather perpetuated the family passion for books, supplementing his portion of his father's library with his own substantial purchases. By Gu's time, this collection had grown to five or six thousand volumes.[14]

The point is that in the sixteenth century it had become possible to own books on a scale surpassing all earlier times. No one in the Song dynasty could reasonably hope to own ten thousand fascicles. By the end of the sixteenth century, dozens of private libraries held ten thousand separate titles, each of which could run to many fascicles.[15] More books were available, and more people read and owned more books, in the late Ming than at any earlier time in history, anywhere in the world. It created something of a mania for collecting. Wang Wenlu was one of those with a library of ten thousand books. He invested an extraordinary amount of not just his money but his energy and passion into the monumental effort of amassing the collection. When his library caught fire in 1568, Wang cried out, "Rewards only for those who go in and save the books, nothing else!"[16]

Gu Yanwu was from a book-collecting family whose scholarship was many generations deep. Li Rihua was not. His collecting instincts drew him more to aesthetic than scholarly objects. He did not go out of his way, as Gu's ancestors did, to snag a title he had been searching for, though he was not averse to taking note of a fine edition when one came his way. So for example when a neighbor brought him a Song edition of the massive thousand-volume imperial encyclopedia of the tenth century, the *Texts of the Taiping Era for the Imperial Gaze (Taiping yulan),* he noted it in his diary. The attraction of this book was that it was an original Song edition, and Song imprints were rare and highly prized. Added to this was the fact that Li knew the man who once owned it, which increased its value in his eyes. He was also aware of the book's financial value, estimating it to be worth the astronomical sum of a hundred taels of silver. He didn't buy it, though.[17]

Li may not have been an avid book collector, but he did haunt bookstores. He recounts going into one in Suzhou and being shown a curious manuscript, the Hongzhi palace edition of an illustrated pharmacopoeia, forty fascicles in four covers (the equivalent of one "title" for a bibliophile hoping to collect ten thousand). Li is impressed. "No previous reign paid this sort of attention to pharmacology," he notes. "It is truly the production of a prosperous age." The bookstore owner explained to Li that he got the book from someone in Wujiang county, just south of Suzhou, and that owner got it from an attendant in the palace—"certain evidence," Li declares, "that a lot of things leak out of the palace vaults." The book was produced under the active patronage of the Hongzhi emperor, but it suffered the misfortune of being completed just before he died. The Zhengde emperor succeeded him, and the manuscript was left to languish in the palace library. One handwritten copy survives today.[18] Was it the copy Li saw?

A manuscript smuggled out of the palace was a curious rarity, which explains why Li was interested in it. Ordinary readers would not have been, especially at the price the bookseller was asking. Most bookbuyers in the Ming occupied a lower stratum of the market, where literacy was a tool of business and pleasure, not of scholarship. But even illiterates seemed to buy a book or two, perhaps just for the social cachet of owning them.[19] More surprising, perhaps, is that complete illiterates may well have been a minority in the late Ming. The Spanish Jesuit Adriano de las Cortes, though unimpressed with the general standard of living he found in the Ming when he was washed ashore in 1625, was impressed by their education. "It is rare," he asserts in his memoir, "that a boy, even the son of a Chinese very poor and of low condition, does not learn at least to read and to write their characters." Being from a culture in which not even all aristocrats bothered to learn how to read, Las Cortes was also struck to discover that "among the great, of whatever quality, rare is he who does not know how to read or write." He found far fewer literate women, due to the general exclusion of girls from village schools. "In all the schools that we entered, we saw only two girls who were learning."[20] Girls who learned to read had to acquire the skill at home, usually from literate mothers, occasionally from fathers or brothers.

The increase in demand pushed the publishing industry toward streamlining and standardization.[21] One effect of standardization was largely to abandon the use of movable type, which in a language of thousands of characters posed problems that typesetting in a script with twenty-six letters did not. Another was to make books even cheaper. As a result, readers could afford to read for pleasure, not just for work. The popularity of extended works of prose fiction in the second half of the sixteenth century—narrative works antedating the novels that would begin to appear in Europe a century or more later—must be attributed at least in part to the development of commercial publishing, for within Li's lifetime appear three of the greatest premodern novels: a tale of marsh outlaws, *Water Margin (Shuihu zhuan,* translated as *All Men are Brothers*); a fantasy adventure, *Journey to the West (Xiyou ji,* translated as *Monkey*); and an erotic novel of merchant life, *Plum in the Golden Vase (Jinping mei).*[22] We know that Li owned a copy of *Water Margin,* but he drew the line at erotica. When commonplace book author Shen Defu (we have cited passages from his *Gleanings from the Wanli Era*) sent Li his copy of *Plum in the Golden Vase* through his nephew, Li declined to accept the book. It was, he stated, "on the whole an extremely filthy book from the gutter press, far inferior in wit and impact to *Water Margin.*"

The lowering of cost also meant not just the spread of hugely popular books but the coming into print of bodies of specialized knowledge, such as medicine, that previously might have never left the manuscript stage. We know that Li Rihua owned medical books, for he mentions "parsing and reading medical books, and gaining some insights" while tending his sick wife in 1613. The reference to "parsing" has to do with the convention in Chinese publishing of not supplying punctuation marks. The text simply unrolls character by character without any marks to show where a sentence begins and ends. Context and the use of certain "empty" characters provided readers with sufficient clues as to where to break sentences, though reading a text on an unfamiliar subject could pose some challenges. "Parsing" in fact is an overtranslation of the word Li actually used, which was *dian* or "dotting." This refers to the practice of adding dots of ink at the end of each phrase, which provided rough punctuation and marked one's place in the text. Many readers liked to "dot" as they read.

In addition to medical texts and novels, Li mentions in his diary buying scholarly writings. On October 14, 1611, a friend brings him a newly engraved edition of the massive compendium—124 fascicles in ten covers—of historical writings on state organization edited by the great scholar Tang Shunzhi (1507–1560). The book survives today in a 1595 edition published by the Southern Academy in Nanjing, where Tang had been chancellor.[23] Li already owned two older editions of the book, one of them probably the Southern Academy edition, so what the friend brought must have been a later commercial copy. Li clearly prefers the new edition, for when a traveling merchant from Huzhou comes by later that same day, he trades the two in for several books the dealer had on hand, including a Songjiang imprint of a popular collection of historical anecdotes.

The center of commercial publishing during the Yuan and Ming was Jianning prefecture, deep in the mountainous interior of Fujian, particularly the town of Jianyang.[24] Li Rihua owned Fujian imprints. On February 2, 1610, he notes that he received a package from a student in Fujian that contained two bottles of liquor, four preserved oranges, a catty of tea, and a newly engraved edition of *Annotated Record of the Scrutiny of Craftsmen* (*Kaogong ji shuzhu*). This was a chapter of the Han-dynasty classic, the *Rituals of Zhou* (*Zhou li*), dealing with manufacturing to meet the needs of the central state. The larger classic was a great favorite of Ming scholars, who used it to imagine the state as it should be run rather than as it actually was. The chapter on craftsmen was famously obscure, so that the author of this edition took up the pedantic task of explaining everything in the original. The only surviving edition of this book, published in 1603, is roughly engraved and copiously illustrated, both of which are signs of a Jianyang imprint, as Li's seems to have been.

Li's personal tastes did not go to the Jianyang end of the market, however. The books he most frequently notes as his reading matter in the diary are Buddhist sutras. He had a particular attachment to the *Flower Garland Sutra* or *Avatamsaka Sutra*, the most authoritative scripture of the Mahayana sect of Buddhism in China.[25] On July 10, 1610, a friend showed him a Song edition of the *Flower Garland Sutra* dated 1092, with an inscription inside the back cover by a famous monk, which only added to the value of the book as a collector's item. On December 16, 1612, Li received a copy of *The Garland Sutra and Exegesis in a New Edition* (*Huayan xin jinglun*) from Pan Zhiheng, a prominent painter and biographer from Huizhou. The original commentary was by a Tang author, and a Buddhist monk had combined sutra and commentary together to produce a joint edition. Li explains in his diary that Pan has extracted the sutra from this edition and published it separately with a preface by Jiao Hong (1541–1620), the eminent Nanjing Neo-Confucian who advocated the Unity of the Three Teachings. As it happens, Jiao also contributed a preface to the edition of Tang Shunzhi's compendium published by the Southern Academy, of which he was chancellor at the time. Li may have traded in his copy when he was able to replace it with a new edition.

Four months later, Li notes that he has just returned home after ten days on his boat, and that while boating he "parsed and read *Combined Commentaries on the Garland Sutra* (*Huayan helun*)." He adds that "the principles of the Buddha are so excellent that it would be the mistake of a lifetime not to read this book." A year later, a friend showed him a copy of the *Garland Sutra* consisting of eighty-one fascicles in sixteen covers. The man took six years to write it out by hand and invited Li to add an inscription at the end of the book. A book such as this was an object to pass around and admire, an occasion to engage with friends and share commitments—elegant consumption, even if the object being consumed was a religious text that showed the path to non-attachment.

Furniture

Li Rihua took pleasure in furniture styled in a way that is still with us, as what we call "traditional" Chinese furniture: rosewood chairs in which the arms and back are joined by a single curve of thin wood, the slope-sided cabinets, domed trunks, spindly clothes racks, folding stands for wash basins, and four-poster beds filigreed with fine latticework. The history of any of these pieces may go back before the Ming, but we know them best in their Ming form, for almost no furniture predating the Ming survives today—though most of what is on display in museums as Ming furniture has been rebuilt in the Qing or later and may contain no more than a few pieces of wood that actually date back to the Ming.

Furniture in the Ming became increasingly refined in style and construction. Manufacturers carved their pieces more delicately and made panels and arms curve to fit the contours of the body. Joinery improved, such that the tenons or pins inserted to hold pieces together were no longer exposed to view.[26] Most striking, though, was a shift to hardwood cut to a thinness cheaper woods could not tolerate. Fan Lian (b. 1540), a cultural arbiter of the early Wanli era, registers this shift in his 1593 commonplace book, *Notes on What I've Seen on the Delta (Yunjian jumu chao)*. The taste for fine hardwood came in during his own lifetime, for he declares that such furniture "was not to be seen when I was young." Back then, "people made do with square lacquer tables of gingko wood, but now the wealthy crave the thin hardwood furniture made in Suzhou." The effect was escalation, driving buyers and makers to move to ever more expensive woods such as cherry, ebony, and boxwood, even for everyday items such as beds and cupboards. The lesson? "The utter waste that fashion causes."[27]

Chairs were a particular accomplishment of Ming furniture makers, and they appealed to buyers, to judge from the 61 chairs in the Jesuit house in Nanjing. Chairs were not widely used until the Northern Song dynasty, when the allegedly barbarian practice of lounging on couches (which in turn had superseded an even earlier practice of sitting on mats on the floor) lost favor. Ming furniture boasted a wide variety of chairs, some of which were innovations of the dynasty. But couches were still considered suitable for the gentleman who "in moments of pleasant relaxation would spread out classic or historical texts, examine works of calligraphy or painting, display ancient bronze vessels, dine or take a nap, as the furniture was suitable for all these things." This is a quote from the chapter on furniture that Wen Zhenheng included in the handbook of elegant taste he compiled late in the 1610s, *Treatise on Superfluous Things (Zhangwu zhi)*.[28] Wen does not acknowledge that couches were problematic, being associated in the popular mind with a dissolute second-century emperor famous for lounging on his "barbarian couch." The Ming founder was also hard on couches. After defeating his military rival Chen Youliang, Zhu Yuanzhang's officers presented him with Chen's carved gilded couch. Zhu found the object offensive for the amount of wealth consumed to create it, and ordered it destroyed.[29] Ming rulers would sit on upright chairs, not lounge about—perhaps implying that their subjects should do the same. Clearly the message had faded by Wen Zhenheng's time.

The shipwrecked Las Cortes was impressed by the chairs he saw, "extremely well made and carved, albeit in a barbarian style," which seems to be his way of registering what he regarded as Mongol influence. Las Cortes then goes on to describe the other furniture one would expect to find in a wealthy home. He was particularly amazed by the number of small tables, again "extremely well made," that were to be found in a home, anywhere from twenty to forty in a large public room. They were "stacked one on top of the other and not ordinarily used, except for one or two of them." Las Cortes explains this profusion of stackable small tables, unimaginable to European taste, by noting to the reader that "there it is a form of ostentation."[30] Indeed, they were a fairly recent style. Through the Yuan and early Ming, the standard square dining table, known as the Eight Immortals Table, sat two people to a side for a total of eight. In the sixteenth century, these larger tables gave way to smaller tables seating at most two people, reflecting a concern to maintain distinctions of hierarchy that might otherwise become blurred when eight people sit down to a meal together.[31]

The inventory of the Sun family suggests that Huizhou was behind the times in 1612 when they divided their property. The Suns still had 4 of the old-fashioned Eight Immortals Tables, plus 1 large incense table, 6 tables for musical instruments, 4 lacquer tables, 3 tables with drawers, 6 collapsible round tables, 4 collapsible square tables, and another 8 that are dismissed variously as "old," "small," or "rough."[32] No small stacking tables in sight.

The art historian Craig Clunas has noted the anonymity of Ming furniture makers, observing that "not a single name of a producer is recorded in the writings of the consuming class."[33] But Li Rihua gives us the names of two restorers, at least. Restoration was an admired but ambiguous skill. It could bring ancient pieces back to their original "elegance," and this was a good thing; it could also produce forgeries, not a good thing. Li admires Zhou Danquan, a restorer of wooden objects in Suzhou, calling him "extremely clever. Any broken vessel or stringed instrument that he touched he could restore; anything vulgar he could make stylish. For a time he was valued above everyone in Suzhou."[34] Li suggests that Zhou learned his skills from the Daoist mystics with whom, it was said, he associated.

The Suzhou lacquerer Jin Meinan, by contrast, was not above using his skills to dupe his customers. On September 28, 1615, Li writes that he had just bought a black lacquered couch inlaid with Longtan stone. Longtan, in the north end of Jiangxi province, supplied a veined stone that was considered acceptable but nonetheless inferior to "phoenix stone," which had to be transported a much greater distance from Dali, Yunnan. Li mentions elsewhere buying two screens and two lounge chairs with Dali stone inlays from the barge of a traveling merchant from Wuxi, so we know that he had examples of the real thing. But as Dali stone was expensive and difficult to obtain, Longtan stone emerged in the sixteenth century as an affordable substitute through the ingenuity of Jin Meinan. While visiting Longtan, Jin discovered stone that, "when ground and polished, had a beauty to rival Dali phoenix stone," Li writes. "So he hired laborers to dig up some of the stone and cut it into slabs, then treated and improved its surfaces by following the natural patterns. From it he made screens, tables, chairs, and couches. At a glance, no one can tell that it is not Dali stone." Li caps the story by recounting how Jin used Longtan stone to trick a high official into paying sixty taels of silver for an elaborate piece of lacquer furniture he thought had Dali inlay but was really only Longtan. Though ever on guard against fakes, Li seems to feel that the social-climbing official, who was embarrassed upon discovering that he had been duped, got what he deserved. Here was exactly the sort of buffoon who should be kept out of the zone of elegant possession that Li policed.

Porcelain

Surely the most common expression of Ming style that is still with us today is the object that Europeans in the Ming dynasty named after its place of origin, china. The style that emerged during the Yuan is now classic: thin white porcelain decorated with cobalt blue designs trapped between two layers of glaze fired at temperatures so high that the surface becomes utterly transparent and hard as glass.

Porcelain is a Chinese invention—but the conventional decorative blue designs on a white body are not. This intercultural aesthetic was created by the international pottery market. The taste for blue on white originated in Persia. Persian potters lacked the technical capacity to produce pure porcelain, but they did have cobalt dark enough to mark vivid decorations on the surface. Seeing what Persians liked, Chinese potters followed suit, employing their superior glazing technology to turn out a much finer product that sold well in Persian markets starting in the fourteenth century. The demand for their wares was huge in part because of a local religious constraint. The ban in the Koran on the ostentatious practice of eating off gold or silver plates (Zhu Yuanzhang issued the same ban within his own family) created for blue-and-white an opening among wealthy Persian consumers who wanted to serve their guests on expensive tableware.

The center for porcelain production in the Yuan was the city of Jingdezhen in the interior of Jiangxi province, and it is still the center of porcelain production today. Jingdezhen developed where it did because of the proximity of large deposits of china-stone, which was pulverized and mixed with other ingredients to produce the paste from which ceramic products were molded, then glazed and fired. Although distant from the main commercial cities of the Yangzi valley, Jingdezhen was well enough connected by water to be able to ship its products down to the Yangzi delta profitably.

The Yuan government brought Jingdezhen potters into its state procurement system in 1278 by establishing a bureau there to manage orders for the court. The bureau was expanded in 1292 and 1324, when Jingdezhen was placed directly under the provincial governor, but it was a brief period of supervision, for Jingdezhen slipped under rebel control in 1325 and was only reincorporated into the state system in 1369.[35] Curiously, 1325 is also the date that we use to mark the changeover in production to blue-and-white. We do so on the basis of cargo recovered from a Chinese shipwreck off the coast of Korea, in which wooden tags were found dating the loading or launching of the vessel to June 1, 1325. The cargo included some five thousand Jingdezhen pieces of various glazes, but not one blue-and-white.[36] Within a decade, however, there is not one cargo or cache of Jingdezhen porcelains that does *not* have blue-and-white. The loss of state control over the potters of Jingdezhen may have been decisive in releasing this sea-change of style. Almost overnight, this intercultural crossover swept Chinese and international markets. Potters in production centers throughout the world, from the court of Tamerlane in the fifteenth century to Mexico in the sixteenth to the Delft in the seventeenth, worked at imitating the look and feel of Chinese blue-and-white—and consistently failed.

Li Rihua bought porcelains of all types from many locations—including Jingdezhen. Indeed, his most extended discussion of porcelain deals with a potter in Jingdezhen named Hao Nineteen. "Nineteen is a skilled ceramicist," he notes in his diary on April 1610. "Everything he makes in the style of the Yongle, Xuande, and Chenghua kilns is near-true. As a person he is elegant, enjoys chanting poetry, and likes painting"—this is Li assimilating the best artisan into his elite cultural world. Li recalls meeting Hao in the spring of 1598 while making a purchase of imperial wares from several kilns at Jingdezhen. "Even then his hair was white. I gave him an order to make shallow bowls in a flowing mist style, glazed in a secret color that combines cinnabar and lead, and paid him thirty taels in cash. Then I had to leave and promptly forgot all about the bowls with the flowing mist." He writes this note because he has just received a letter from the potter. "Today the letter arrived telling me that Nineteen completed fifty items, then gave them to Shen Biehe to bring to me, but they have disappeared." Shen was a notorious hustler from Hangzhou who kept tax accounts for one of the princely establishments. "None of the gentry will have anything to do with him," Li declares, then observes with a sardonic shrug, "No surprise that my bowls have flown."[37]

The Arts of the Brush

Li Rihua read books, furnished his home, and drank tea from fine porcelain cups, but the objects that really mattered to him were calligraphy and paintings. Unlike the everyday luxuries with which he surrounded himself, the scrolls, albums, and fans that preserved the art of the brush were the work not of artisans but of educated scholars like himself. They existed on a different plane, skillful to be sure but not mechanically expert, prized in terms of the values they were assumed to express and priced in relation to culturally shaped demand.

Painting and calligraphy were participatory arts for the elite—Li was a decent painter of landscapes and an excellent calligrapher—but the works that mattered to collectors came from a tiny handful of celebrated artists. This is where Dealer Xia comes into the story. Li could acquire a certain amount of art through his personal networks, but many works existed beyond those networks, and the only way he could even know about them, when they came onto the market, was by collaborating with a dealer who traveled the length and breadth of the Yangzi delta working his commercial networks to turn up fine art that Li might never hear about. Xia did a great deal more than the bidding of his prime customer, however. He also brought into his reach an endless trove of fakes. Li understood that this was part of the bargain. His diary entries suggest that he found detecting fakes to be almost as much fun as discovering real works.[38]

By the turn of the seventeenth century, works of the Tang and Song masters were almost unobtainable: so few survived, and so many wanted those that did. Hangzhou collectors in the early Yuan could still hope to acquire works by, say, the greatest calligrapher of the Song dynasty, Mi Fu (1051–1107).[39] A collector like Li Rihua might still dream of getting his hands on a scrap by Mi Fu, whose style he adored, but the ever widening distance between the mid-Song and the late Ming meant that the odds were severely against him. The available Mi Fus were almost all complete fakes. All Li could do after a visit from a low-class dealer who faked a two-character inscription by Mi Fu himself is splutter, "Hopeless! Hopeless!"

As earlier dynasties stretched away from them, Ming collectors had to turn their attention to Yuan and Ming productions. Among Yuan artists, Zhao Mengfu (1254–1322) was far and away his first choice. He also favored Huang Gongwang (1269–1354), Ni Zan (ca. 1301–1374), Wu Zhen (1280–1354), and Wang Meng (1308–1385), roughly in that order. The certainty of his taste was matched by the fervor with which dealers came to him with paintings and calligraphy that they insisted were Yuan and that he could detect as fakes. When Dealer Xia comes by on December 26, 1609, with a job lot from a prominent Shanghai family, Li is hopeful that works held for generations in an old family should include something really good. He is disappointed. Some of the Ming works are genuine, but he dismisses the Yuan pieces with such comments as "not up to standard," "suspect," and "unreliable." Did the Shanghai family dupe Xia, or was Xia hoping to make a killing from Li? We don't know. But hope springs eternal in the heart of both dealer and collector, and five days later Xia returns with a colleague who is offering a genuine Yuan-dynasty masterpiece by Ni Zan. "The brush strokes are unsurpassedly fine," Li declares delightedly.[40]

Yuan works were already so rare that even a collector such as Li ended up mostly with works by Ming masters. Li favored Shen Zhou (1427–1509), Tang Yin (1470–1524), Chen Chun (1483–1544), and Wen Boren (1502–1575), but his hands-down favorite was Wen Zhengming (1470–1559), who as it happens was the grandfather of Wen Zhenheng, the author of the guide to all things elegant, *Treatise on Superfluous Things.* We have already met Wen as the painter of *Heavy Snow in the Mountain Passes,* which he painted at the start of the 1530s just before the long warm phase in the mid-sixteenth century. Until taste shifted around the turn of the seventeenth century, Wen Zhengming was universally regarded as the greatest painter and calligrapher of his dynasty, as Zhao Mengfu was of his. Li was always happy when a Wen Zhengming came his way. The first diary entry in which he mentions Dealer Xia, in fact, has Xia bringing him a sketch by Wen Zhengming. Li notes that the technique in this painting is "coarse" but likes it nonetheless.[41]

Wen Zhengming would be surpassed during Li's lifetime by only one artist, the painter, calligrapher, and art theorist Dong Qichang (1555–1636). Dong hailed from the adjacent prefecture to the north. Ten years older than Li, Dong perched himself at the front edge of a wave of taste that carried Li and his generation forward. Dong was largely responsible for establishing the celebrated lineage of accomplished scholar-amateurs as the highest aesthetic expression of cultural and moral status, far superior to the mere artisans who painted for the court or for the market. This lineage of gifted amateurs stretched back to the inimitable Mi Fu in the Song, was enriched by the "four great masters" of the Yuan (Huang Gongwang, Wu Zhen, Ni Zan, and Wang Meng), found sublime expression in the paintings of Wen Zhengming in the sixteenth century, and then culminated in none other than Dong himself.[42] Li lived in Dong's intellectual and aesthetic shadow, sharing his tastes and collecting his paintings and calligraphy. Dong's success had as much to do with his position as the chief interpreter of the history of painting as it did with his skill with a brush, though this too was considerable. Our perception of what constitutes "Chinese art" today derives from Dong Qichang.

Market and Taste

Did Li Rihua actually like the objects he acquired? To judge from his comments in his diary, yes, he did. But what he himself felt is beyond our retrieval. What matters is that he made his choices in conformity to prevailing taste. More important, the fact that he could acquire his collection is evidence of the existence of a market that made these objects commercially available. Taste mattered to the acquisition of cultural works, but the market mattered even more. Art was business. Had matters been otherwise, there would have been precious little for Li to collect. The business of acquiring fine things in the Wanli era had changed since Jiajing even, when paintings still moved more within networks of friendship and reciprocal obligation, where they were exchanged as "elegant debts" among a tighter elite circle.[43] Patrons in that era commissioned painters to produce work and paid them well for it. Less often were luxury artworks marketed to anonymous buyers who gained access to them via commercial dealers.

The commercialization of culture did not necessarily alter taste; in fact, it probably reinforced it. But commercialization did alter the effect of demand, which was to draw forgeries into the market in far greater numbers than the genuine articles, turning the work of acquiring things into the much more elaborate game of discriminating between the true and the false. When Dealer Xia drops in on January 8, 1613, with only a few sundry curios and no art, he is dispirited. "In recent days the market in calligraphy and painting has collapsed," he explains to Li. "Sellers won't sell, and buyers won't buy. It is because the fakes are so numerous and those who have been cheated are not few. Everyone is cautious about getting burned, and no one dares stick out his head in this situation." Xia had discovered something like Gresham's law of luxury consumption: objects overvalued by the market (unidentified fakes) will drive objects undervalued by the market (authentic pieces) out of circulation. This was the risk of trading in luxuries, and the peril of collecting them.

For Li, the culprit is clear. It is not his dealer but the mass of uninformed buyers crowding the market. When Xia shows up on March 6,

1613, with a five-inch-long piece of jade he claims is an ancient wrist rest for calligraphers, Li knows it is nothing but a broken piece of unearthed jade and nothing more, and tells Xia as much. He concludes his diary entry on the visit with this homily: "Ever since gentlemen began collecting antiquities in order to satisfy their desire to acquire things they like, young dandies have followed in their wake, not stinting on cash or wealth to buy them. And so crafty dealers and market hustlers will do anything to make a sale, spouting absurd claims to the point of talking complete nonsense." Li regarded ownership of the right cultural things as a sign of good breeding and education. The merely wealthy, having no real understanding of cultural objects, spoiled the market for the true connoisseurs. The hard truth of luxury collecting, of course, was that it required money more than anything else, much to the annoyance of collectors who wanted to distinguish mere acquisition from true appreciation—and so distinguish themselves as disinterested bearers of the cultural tradition, not as rich people looking for investments or social status.

The commercialism of the relationship between Li Rihua and Dealer Xia peaked on November 19, 1614, when Xia showed up with Wen Zhengming's *Picture of Preserving the Chrysanthemums*. Li notes that Xia "was very proud of himself" for having come up with such a fine Wen Zhengming. "I said nothing for a long time, then in a leisurely fashion produced the real one from my collection for his inspection, at which he fled without ado. When I placed the real one beside his, he panicked! Laughable. I bought this scroll some twenty years ago!"[44]

We will never know Dealer Xia's side of the story. Li had the upper hand in determining standards of taste, but Xia had the upper hand in terms of supply. Xia had to keep his supply up if he wanted to stay in business, however, hence his willingness to show fakes to Li and see whether they would pass muster. The market thus supplied Li with the objects he desired and was ready at every turn to trick him into paying huge sums for objects that were worthless. "Nothing but a recently fabricated oddity by someone in Suzhou," he complains of an eleventh-century scroll that Xia brought to him to authenticate—and sell if it proved to be authentic.[45] We can hardly blame the artisans of Suzhou. They were simply supplying the market with whatever would sell. There weren't enough Song paintings to go around, and most consumers had to make do with Suzhou knock-offs. Looked at dispassionately, fakes are proof that supply responds to demand. They are luxury commodities in their purest, if most worthless, form.

By keeping up the demand for certain artists over others and policing their oeuvres by driving out fakes, Li Rihua was doing more than just satisfying his personal need to possess tasteful objects. He was fixing and perpetuating standards of taste that still prevail today as China's "national style."[46] Indeed, the paintings of Yuan and Ming masters might well have been frittered away by time, unrecognized and uncollected, were it not for buyers such as Li Rihua who invested endless time, energy, and money in collecting what mattered to them. The unintended effect of his efforts was to define, not just for his generation but for every generation since, what everyone recognizes immediately as Chinese painting, so too porcelain, and furniture. Less so books, though Chinese book designers today still imitate certain features of the old layout. But however much the artists, artisans, and connoisseurs of the Yuan and Ming have defined Chinese culture as it is today, they did not do it alone. Suzhou commercial artists, Jingdezhen potters, and men like Dealer Xia all played their necessary parts by pursuing the business of things.

Chapter 9 Q's

connection between tribute & trade

silver → development of south china sea → world economy currency

impact of priests

9

THE SOUTH CHINA SEA

GUAN FANGZHOU went to sea. This was not something he would ever have imagined doing earlier in life. He was a successful silversmith with a flourishing business in Suzhou, made more so by lucrative government commissions. This was in the late 1570s, when Chief Grand Secretary Zhang Juzheng was reforming the entire fiscal system by cashing out the old labor levies and converting them into silver payments. Silver had become the currency of the age, literally as well as figuratively. For a silversmith, it was a good time to be in the business, and Guan had become a wealthy man.

Guan had no obvious ties to the sea, but he would have known merchants who did. Though not itself a seaport, Suzhou was the commercial pivot of the entire network of land and sea trade radiating from the Yangzi delta. Wholesale merchants handling bulk exports assembled their cargoes here and then barged them down to the tidal harbors in Taicang, Shanghai, and Jiaxing, where they stowed them on the cargo ships heading down the coast or out to Japan. The lifting in 1567 of the ban that had crippled maritime trade for four decades gave the export business a huge boost. Trade to Japan was still under interdiction, but there was little difficulty in fudging cargo destinations with the customs officials. Guan would have been indifferent to the silks and ceramics going out. But he would have been watching with intense interest the arrival of the highly valued material captains were bringing back from every foreign trade entrepot around the South China Sea: his stock in trade, silver. Lifting the coastal ban meant that silver was entering the country at a far greater rate than the legal mines in Yunnan and the illegal ones scattered

throughout the south were producing bullion. The Jiangnan economy was flush with cash, as Zhang Juzheng well understood when he pushed the tax system onto the silver standard.

Guan would never have had to go to sea if he had not been found out. Everyone knew that a sharp silversmith had a hundred ways to trim the silver that came into his possession to his own advantage. Guan went too far, defrauding the government to the tune of a thousand ounces. When his embezzlement was exposed, he was thrown in the judicial prison of the Suzhou Guard to await a directive from Beijing on punishment. Should Guan be found guilty under article 487 in the Ministry of Works section of the Ming Code dealing with ordering more materials for a project than was needed? If so, that statute referred the judge to the scale of penalties in article 287 in the Ministry of Justice section dealing with custodians stealing money from state treasuries: if the theft was valued at the equivalent of forty ounces of silver, the penalty was decapitation. Or should the penalty be taken from the next article (288) on ordinary people stealing money from state treasuries? If so, the theft had to reach eighty taels before the penalty was execution, in this case by strangulation. As Guan's theft went well above that threshold as well, the only question was which form of execution should be applied. Article 487 imposed decapitation, whereas article 288 demanded strangulation. Strangulation was preferred, as it left the body intact and posthumously able to receive sacrifices and achieve rebirth. Guan enjoyed a temporary reprieve while local officials awaited the authorization.[1]

The warden of the prison, surnamed Wang, was related to Guan's son by marriage. It was a connection Guan was quick to capitalize on. Warden Wang was known to be an easy-going jailer in any case. Guan soon worked out an arrangement that allowed him to come and go from the prison pretty much as he pleased, so long as he was back behind a locked door every evening. Then came the day when Guan left the prison and failed to return by sundown. The imperial censor exploded when he heard of the escape and ordered that Warden Wang should bear whatever punishment the Ministry of Justice sent down for Guan on the principle that the crime had to be punished even if the original criminal was unavailable. The pressure was now on to find Guan before the sentence was carried out. Wang's family spent a fortune sending out spies all over the region to track Guan down.

There was widespread sympathy with the warden's plight, so a local official ordered his constables to join in the hunt. It was to no avail. The only scent they picked up led to the sea. It seems that Guan had found his way onto a ship. The constables' best guess was that he had sailed down the coast, so they went in the same direction, scouring the ports of Fujian and Guangdong. Despite their best efforts, they came up empty-handed. Guan had disappeared without a trace.[2] But this, as we shall see, was not the end of the case.

Artifacts of Maritime Trade

At one end of the Eurasian continent, a silversmith slipped down a canal to a port on the south bank of the Yangzi River estuary, got on a ship, and headed out to sea. At the other end of the continent, a library in a town up the Thames River from the port of London took delivery of its first Chinese book. In the larger scheme of things, it was a precocious acquisition. The year was 1604. London's first shipment of Chinese tea, till then an untasted beverage in Elizabethan England, would not arrive for another five years.

The town to which the book was destined was Oxford. The university library had been founded barely four years earlier, the late-career project of a retired civil servant named Thomas Bodley (1545–1613). His timing was apposite. Gutenberg had adapted the Chinese technology of movable type printing less than a century earlier, and the uptake had been swift. Universal knowledge once required reading a limited number of books. Now that number was unlimited. No one person could own copies of everything. Scholars had to band together; common libraries were now necessary. Bodley had a particular interest in books published in what were then called "modern languages" to distinguish them from the classical languages of Greek and Latin. As a young man, Bodley recounts, he was "desirous to Travel beyond the Seas, for attaining to the Knowledge of some special Modern Tongues" and had spent four years in Italy, France, and Germany doing just that. He understood that his library's mission should be to acquire books and manuscripts in all languages, not just those he knew but other languages as well.[3] And so it was in 1604 that Bodley took receipt of his first Chinese book.

Bodley probably acquired it through one of his purchasing agents in Amsterdam, who in turn bought it from someone in the Dutch East India Company. The VOC (Verenigde Oostindische Compagnie) had been incorporated just two years earlier. Its creation was a masterstroke of the Estates General, the fledgling Dutch government, which forced the first generation of Dutch merchants trading in Asia to compete with the Portuguese and the Spanish rather than with one another. Within a decade, Amsterdam replaced Lisbon as the point of arrival for goods coming from the East. Books were rare objects among the early goods offloaded onto Amsterdam's wharves. They were mere curiosities, as no one in Holland or England could read Chinese. But Bodley saw the point of buying them, for one day, he was sure, someone could unlock the knowledge they contained.

Chinese books were random acquisitions in Bodley's day. In 1635 the Bodleian Library, as it became known, received a few volumes from its first major gift of Asian books, one of several donations from William Laud (1573–1645), Archbishop of Canterbury and Chancellor of the University. Laud liked to collect a range of Asian books and manuscripts, he too being confident that English scholars would eventually master these languages. Three years earlier, Cambridge had appointed its first professor of Arabic; a year after the gift, Laud himself installed Oxford's first Arabicist. Laud's donation the following year included one more Chinese book.[4] Most of Laud's Chinese books are ordinary commercial printings of novels and primers, the sorts of books one might expect a sea captain's family to own as pleasure reading or instruction for his children, not what a Ming scholar would acquire for his library. But there is one book Laud donated that is more precious to historians than what the greatest Ming scholarly libraries owned. Presented to him in 1639 by a visiting Jesuit, it is a manuscript copy of a rutter, a navigator's guide (in words, not in maps) to sea routes connecting China to the world. Catalogued by the handwritten title on its cover, *Dispatched on Favorable Winds (Shunfeng xiangsong),* it is known today as the Laud rutter.[5] Starting from the southern coast of Fujian, the rutter gives compass bearings for routes out to Ryukyu (Okinawa) and thence to Japan, to the Spanish port of Manila in the Philippines, down to Brunei, around Southeast Asia, and to ports in the Indian Ocean, principally Calicut (in the present-day state of Kerala), and from there to Hormuz at the mouth of the Persian Gulf. Derived at least in part from records of the voyages of Yongle's eunuch, Zheng He, it is unique.

The historian Xiang Da visited Oxford and prepared a modern edition of the Laud rutter published in 1961, yet it had little impact on how historians of that xenophobic era wrote Ming history. The rutter was treated as evidence that some Chinese went to sea, but it did not alter the prevailing understanding of Ming China as an agrarian empire indifferent to the rest of the world. In fact, the rutter tells a far more dramatic story that not only puts the people of the Ming on the ocean, but shows them actively engaged in weaving the threads of commercial webs that were tying the Ming to the rest of the world, and by so doing, creating the conditions for the rise of capitalist enterprise in Europe.

We now tell a different story about the Ming in the world, and the Bodleian Library is again supplying the evidence, this time in the form of a map donated to Oxford by John Selden (1584–1654). In addition to being a successful lawyer in London, Selden was Oxford's first scholar of rabbinic studies. His work on Hebraic law and Semitic mythology attracted the attention of many, including the poet John Milton.[6] In addition to being Oxford's first Orientalist, in the scholarly sense of the word, Selden was also a fervent advocate of what he called "the rights and privileges of the subject." His particular target was King Charles. Selden's attack in 1629 on royal import duties, which he regarded as an arbitrary abuse of power, sent him to Marshalsea Prison—from which none other than Archbishop Laud, who admired his scholarship though not his politics, secured his release the following year.[7] Selden championed the same issue in the Long Parliament of 1640, to which he was returned as the member for Oxford. The second of the declarations the Long Parliamentarians drew up in December 1640 may even betray Selden's voice: "that the king hath not power to lay any imposition upon forrayne (much lesse homeland) commodityes without Consent of Parliament."

Selden bequeathed his library, which included Oriental manuscripts, to the Bodleian. One of these manuscripts is a large wall map (Fig. 17).[8] Nothing like it exists in any other version or copy. Place names used on the map show it to be from the Ming (it shows the Ming province of Huguang, not the Qing provinces of Hubei and Hunan), but it is not really a map of the Ming. The Ming realm is jammed into the top two-thirds of the map, its northern half oddly truncated and distorted. The cartographer's real subject is maritime commerce, for he has traced a web of lines connecting one point off the coast of Fujian to all the other named places around the South China Sea. Wherever a route shifts direction, he has inscribed the compass bearings a mariner must use to reset his course. The map extends only as far west as the Bay of Bengal, but a cartouche over Kerala gives directions to Aden, Djofar, and Hormuz—all of which were visited by the eunuch admiral Zheng He.

The Selden map fits to the Laud rutter perfectly: the glove of cartography to the hand of the written text. The fit is at one level purely acci-

Fig. 17 The Selden map. This unofficial seventeenth-century wall map, donated by John Selden, depicts East Asia from Siberia in the north to Java in the south, and from Japan and the Philippines in the east to Burma in the west. The Bodleian Library, University of Oxford.

dental, as the two objects arrived in Oxford from completely different sources. At another level, though, they converged at a key moment in the history they document: the linking up of European and Chinese trade.

Tribute and Trade

Since the Han dynasty, Chinese regimes have organized their relationships with foreign states through two mechanisms, one formal and the other only partly formalized: tribute and trade. The tribute system required foreign rulers to send embassies to China bearing tribute in the form of local exotica. The emperor in turn presented these emissaries with gifts of equal or greater value, which they took back to their rulers. The emperor also bestowed titles on tributary rulers, and might even name his favorite in a succession dispute. It was a device for mutual recognition and mutual legitimation that propped up China's claim to world hegemony. Though a fiction, it was one in which both sides happily participated. It gave China the international status it craved, and other states the opportunity to trade.

Even by the more relaxed standards of the Song dynasty, the Yuan was easy on matters of tribute and trade. Khubilai closed trade with Japan in order to prevent Chinese merchants from supplying the Japanese, with whom he was at war, but his ambition to dominate Southeast Asia sent Chinese fleets in that direction, and Chinese traders in their wake. As early as 1277, the Yuan set up four maritime superintendencies at Shanghai, Hangzhou, Ningbo, and Quanzhou. The three northerly offices were concerned with monitoring trade with Japan, though mariners on the Yangzi delta were soon building huge ships and sending them to Ryukyu, Vietnam, and Malacca as well as to Japan. Shanghai prospered so well that the court gave it county status in 1290. The southernmost office in Quanzhou focused on trade into the South China Sea, this being the most important trading port for maritime Muslim merchants from abroad.

The Yuan state imposed a government monopoly on oceanic voyages in 1284 in the hope of generating revenue, but relaxed it a year later, presumably finding that its capacity to manage maritime trade was not equal to private commerce. The imposition of a complete ban on overseas trade in 1303 was the beginning of a serious strangling of the coastal economy. The ban was lifted in 1307 for four years, reimposed, then lifted in 1314, though this time only for official voyages. The last ban in 1320 was lifted in 1322, and until the end of the dynasty, private merchants were free to trade. One effect of the open coast was that the economy of Quanzhou in particular fell increasingly under the control of foreign merchants. Another was that the concentration of wealth in the port cities, rather than bringing prosperity to their hinterlands, undermined it, eventually driving the Fujian coast into rebellion in 1357.[9]

Few records commemorate any of this trade. One that does is a Yuan map that survives today only in Korean versions. The *Universal Map of the Frontiers* (Korean: *Honil kangnido;* Chinese: *Hunyi jiangli tu*) was drawn in 1402 on the basis of a map a Korean diplomat acquired while on a mission to the Ming three years earlier. This map is attributed to Qingjun, none other than the Buddhist master who officiated at Hongwu's plenary mass for the war dead in Nanjing in 1372. The only map by Qingjun that survives in Chinese sources, dated to 1360, stretches west only as far as Burma, though a cartouche off the southeast coast notes that the sea journey "from Quanzhou to Java takes sixty days, to Malabar one hundred and twenty-eight days, and to Hormuz over two hundred days."[10] Qingjun's map bears the title *Broad-Wheel Map of the Frontier Regions (Guanglun jiangyu tu)*. To it the Korean cartographer has added Korea on the right, hugely enlarged, and the rest of Asia and Africa on the left: an oddly elongated Saudi Peninsula, a shrunken Africa, and a clearly recognizable Mediterranean and Black Sea, based presumably on an Arab source.[11] The map is evidence that Chinese had wider knowledge of the world in the Yuan and early Ming than was once supposed.

The Hongwu emperor cared deeply about receiving tribute embassies. Every visit confirmed his right to rule, to potentates beyond his borders as well as to his subjects watching the foreign embassies enter the capital. No tribute missions arrived in his first year, but in his second he received tribute from Champa (southern Vietnam), Annam (northern Vietnam, known after 1428 as Dai Viêt), and Korea. In 1370 Champa again sent tribute, but so too did Java and the Western Sea, that is, Coromandel on the southeast coast of India. In 1371, Annam and Korean emissaries returned, but added to the list were ambassadors from Borneo, Srivijaya (Sumatra), Siam, Japan, and Cambodia. In 1372 the states paying tribute grew to include Suoli, Ryukyu, and Tibet. Hongwu was gratified by these missions, and late in his life was content to look back to the early years of his reign and recall, with some slight exaggeration, that "envoys came continually." He was also on high alert to every slight and shortfall. He rejected the Korean mission that showed up in 1379 with a gift of a hundred catties of gold and ten thousand ounces of silver, which far exceeded what protocol required. In the following year, 1380, he rejected the Japanese mission on the grounds that it did not carry the correct documentation. Japanese feudal lords competed with one another over the right to send tribute missions, and one must have stepped in to preempt whoever had authorization.[12] This was the year when things went terribly wrong, and all because of the tribute system. When the embassy from Champa arrived, it was the prime minister, Hu Weiyong, who received them, not the emperor. Diplomatic theater it may have been for the tribute bearers, but for the emperor this was deadly serious politics.

The Yongle emperor looked to the tribute system for the same reassurance. The *History of the Ming* reports no tribute missions during the unsettled four years of the Jianwen era, but in 1403, once Yongle was on the throne, most of the usual states resumed sending tribute.[13] Yongle exceeded his father, however, by sending his Muslim eunuch Zheng He on expeditions to China's tributaries throughout the ocean they called the Western Sea, and we call the Indian Ocean. If the tribute system provides the framework for understanding these voyages, as we have seen, it also helps to explain their cancellation, for once it was fully functioning, the system did not require the extravagant return missions that Yongle had been sending. Although the expeditions were shut down, the knowledge they had acquired still circulated in Ming society, in the Laud rutter and Selden map, for instance, and in popular late-Ming encyclopedias.[14]

The association between tribute and maritime travel remained strong to the end of the Ming dynasty. The dragons agreed. While crossing to Ryukyu, an envoy from the Wanli emperor encountered not one dragon but three. "We were halfway there when a typhoon arose," writes Xie Zhaozhe, the grandson of the Fujian official who made the local arrangements for the envoy's travel and sailed with the mission. "Thunder, lightning, rain, and hailstones fell upon us all at once. There were three dragons suspended upside down to the fore and the aft of the ship. Their whiskers were entwined with the waters of the sea and penetrated the clouds. The horns on their heads were visible but below their waists nothing could be seen. Those in the ship were in a state of agitation and without any plan of action."

An experienced mariner on board came up with a way to understand the sighting. "This is no more than the dragons coming to pay court to the commissioner's document bearing the imperial seal," he insisted. Xie continues: "He made those attending on the envoy have the latter write a document in his own hand declaring that the court audience ended at such-and-such a time. The dragons complied and withdrew at the time so indicated." Xie draws the necessary conclusion from the sighting: "the Son of Heaven has effective authority over the manifold spirits. It is a principle that cannot be doubted."[15]

The Coast as Border

Tribute and trade were able to sustain each other so long as state diplomacy and foreign trade did not run into conflict. When they did, it was usually because smuggling was placing pressure on state control—and customs duties. Emptying the coast was one response the Ming could take. Hongwu, for instance, ordered the coastal residents of Zhejiang to move inland in order to starve the Japanese smuggler/pirates who raided the coast, a diplomatic move that had heavy consequences for private trade. Similarly concerned, Jianwen forbade coastal residents from having private contacts with foreigners or from warehousing or selling foreign goods.[16]

The other situation that could put trade and tribute in conflict was the arrival of traders claiming ambassadorial status, which not only offended the Ming but could have serious political consequences back home. In 1493 while serving as supreme commander of Guangdong and Guangxi, Min Gui (1430–1511) appealed to the Hongzhi emperor to do something about the huge number of foreign ships landing in China without reporting their arrival to officials and without any regard for the tribute schedule, even when they happened to be authorized as tributaries. Min was not hostile to foreign trade; he was merely trying to address two budgetary concerns: the collapse of customs revenues and the expense of watching the entire Guangdong coast. He asked the emperor to issue a strong notice stressing the inviolability of the rules for submitting tribute. The emperor forwarded Min's request to the Ministry of Rites for an opinion.

In his response, the minister is about as unenthusiastic as he could be without actually suggesting the tribute system be abandoned. True, a lax border policy will just encourage more ships to arrive, but an overly strict policy could strangle the flow, and that would entail an economic loss for the region. He gently reminds the throne that "cherishing men from afar," the elliptical slogan for keeping foreigners at arm's length, should go hand in hand with providing a sufficiency for the country; in other words, let the trade continue. Issuing a stern proclamation would just injure foreign relations and erode trade profits. The emperor should do nothing. Hongzhi agreed, though so as not to discourage Min for his vigilance, he came up with a split decision by promoting him to Minister of Justice the following year.[17]

A sure sign that trade was freely seeping around tribute is the surprising appearance of Folangji (Franks) on the short list of tribute bearers in 1520.[18] "Franks" was a term the Arabs had for centuries used for Europeans. The usage had slid east to name the Portuguese, who had recently arrived in Guangzhou and were trying to claim tributary status with the Ming court in the hope of opening trade. The Portuguese had moved into the South China Sea aggressively in the 1510s, their piratical activities driving the entire regional trading economy into a slump. From then right through the 1520s, Ryukyu—east of China and well out of Portuguese reach—was the only overseas state that submitted tribute with any regularity. The Portuguese bid to be recognized as a tributary was an attempt to gain entry to China in order to trade, and use that access to dominate trade all around the South China Sea. They did not succeed, but the disruption was sufficient to paralyze trade by others. Coinciding with attempts by feudal lords in Japan to force China into trade, the violence catalyzed anti-trade opinion at court. In 1525 the entire coast was shut down. No coastal vessel of two masts or more could put to sea, which excluded everything but small fishing boats. As a popular saying of the time put it, "Not a plank was allowed out to sea."[19]

Closing the maritime border was effective in the short term. The wave of piracy from 1504 to 1524 came to an end. The long-term impact, however, was to promote more piracy by driving traders into smuggling. As competition intensified among the smugglers, they armed themselves, thereby re-escalating violence along the coast. Pirate activity surged in 1548 and stayed high through the late 1550s and 1560s. Several officials won reputations for piracy suppression during these decades, but nothing could change until policy changed, and that had to await the demise of the main author of the coastal ban, the Jiajing emperor. Jiajing finally succumbed in 1566, probably the result of an accumulation of poisons in the longevity drugs his Daoist alchemists were plying him with. As soon as he was dead, requests to lift the ban poured in, as did petitions to improve the infrastructure for maritime trade, including elevating Moon Harbor, Fujian's principal import-export harbor, to county status. The new administration agreed. With the pointed exception of trade with Japan, maritime trade reopened in 1567. Within a year, the Chinese were fully back in the trade. There was one pirate attack near Quanzhou in 1568; thereafter, no major piracy disturbed the coast for the next sixty years.[20]

The ban on trade with Japan soon became a dead letter. Merchants from Canton all the way north to Chongming Island in the mouth of the Yangzi River near Shanghai were sending vessels to Japan and setting up agents there to handle foreign commerce. The scale of this trade can be imagined from the ship that a Jiaxing magistrate seized in the winter of 1642 on the charge of smuggling. It was carrying a cargo of ginseng, probably imported into Japan from Manchuria and then re-exported to China. The magistrate claimed that the cargo was worth a stunning one hundred thousand ounces of silver. The merchants handling the trade, who were not local but hailed from Shanxi province, filed a complaint with the delta's military commanders, hoping to recoup the cargo and stave off the huge loss that confiscation would impose. The magistrate managed to protect his action by giving out lavish gifts of ginseng to his superiors, but was cashiered when a new Grand Coordinator arrived from Beijing and exposed the corruption scheme.[21] The tension between tribute and trade thus marked what was as much a fault line in the relationship between public officials and private merchants as it was a gap between foreigners and the people of the Ming.

To open the coast or close it was a perennial question at court right down to the end of the 1630s. The argument for an open border that Christian convert and Vice-Minister of War Xu Guangqi made in the 1620s was that the Ming needed to have access to the newest improvements in European ballistics technology. His proposals excited angry debate at court. The question was who posed the greater threat, the Europeans and Japanese who came by water to the coast, or the Tungusic warriors, soon to take the name of Manchu, who were pressing on the northern border. Xu had no doubt: it was the Manchus the Ming should fear and prepare against, not the Europeans. Not everyone agreed. His opponents regularly strove to undermine his proposals by accusing him of protecting the Jesuits and selling out Chinese interests to the Portuguese in Macao, so that attempts to borrow European technology and expertise were always compromised and had little cumulative impact on the Ming's defensive posture.[22]

The strongest argument for keeping the coast open, however, was economic. So many people profited from the trade that, as one author commenting on Macao in 1606 gently phrased it, "I'm afraid that in the end trade cannot be banned." By the 1630s, the Ministry of War estimated that a hundred thousand Fujianese shipped out annually to Manila for work. "The seas are the fields of the Fujianese, for the people living along the coast have no other way to make a livelihood," wrote a Fujian petitioner a year after the coast was once again closed in 1638. "The poorest always band together and go to sea to make a living. The moment coastal restrictions are tightened, they have no way to get food, so they turn to plundering the coast. Coastal people must helplessly stand by and watch all they have—their sons and daughters, their silver and goods—taken from them."[23] This was no exaggeration. The ban he was petitioning to remove reduced the number of junks sailing to Manila from 50 in 1637 to 16 in 1638. The collapse rippled through the entire coastal economy. Fortunately for the Fujianese whose livelihoods depended on the trade, the ban was lifted in time for 30 junks to catch the spring winds down to the Philippines—nowhere near the level in 1637 but enough to get trade moving again.

By the late Ming, the decision to raise or lower the barrier on foreign trade was no longer an internal matter. It depended on a host of external factors that interacted with internal concerns. And of these external factors, surely the most important were the changing patterns of global trade.

The South China Sea World-Economy

When Zheng He sailed around the South China Sea and into the Indian Ocean in 1405, he was moving through a zone of existing if dispersed trading networks. When the Portuguese captain Vasco da Gama entered the Indian Ocean in 1498, the same situation prevailed. Muslim merchants based in South Asia dominated the trade, but no one exercised exclusive control. The Zheng expeditions were still remembered around the Indian Ocean when the Portuguese arrived. The memory would have stuck because of the unusual scale of these expeditions, and also because of their unusual character, which seemed to indicate a different mode of operation, that is, a coordinated state-based alternative to the multipolar, segmented system of trade then in existence. The Portuguese were curious to hear that "white-skinned" foreigners—which is how South Asians remembered the Chinese—had once visited all the major ports around the Indian Ocean. As their ambitions in Asia grew, some looked to Zheng He as a model, whether that model was the wisdom of refusing to seize colonial possessions, as some praised Zheng for having done, or the institution of accepting tribute from port states willing to trade with the Portuguese.[24]

The Portuguese arrival in 1511 at Malacca at the western edge of the South China Sea was violent. When they discovered a Chinese commercial community already based there and handling a brisk trade, they decided to treat them as their main competitors and do what European traders as a general practice did to their competitors: kill them and take over their business. This discovery would be repeated over and over again. Wherever Europeans showed up, Chinese were already there. The Portuguese attempted to become a tributary of the Ming, but the Ming rejected the request, as it did all overtures to establish diplomatic or trade relations in order to protect the existing monopoly on maritime trade.

This is why the South China Sea became a critical zone for the eventual integration of the Ming economy with the global economy. The tribute system allowed foreigners to enter China as tribute bearers, but it also required them to exit. Foreign merchants were forbidden from residing in the realm on a permanent basis, and the Ming had the military power to enforce this condition. Anyone who wanted access to the Chinese market, whether to buy or to sell, had to go through state channels and establish a bilateral relationship, the terms of which the Ming always controlled. The only space for private trading was at offshore islands and in smugglers' coves—not a stable foundation for sustained exchange. And so a zone of circulation had to emerge to manage the sale of Chinese commodities leaving China and the foreign imports entering. What emerged around the South China Sea, and what the Portuguese became part of, was a network of multilateral exchanges among merchants tied for the most part to states that submitted tribute to the Ming, but who developed an intra-regional trade in which Chinese manufactures and grain were the leading trade goods.

This trading arrangement rested on one economic condition and one political condition. The economic condition was that the Ming economy had to continue producing goods of sufficient quality and reasonable price to be in huge demand elsewhere: China was the motor of this growth. The political condition was that the Ming state had to continue denying foreign access to its domestic market. Neither condition faltered. Indeed, we could say that the growth of the commercial economy through the sixteenth century, combined with a border-closure policy that only relented in the last third of the century, ensured the strength of this trading system. It was a network sufficiently robust to constitute what may be called a "world-economy."

The term, coined by the historian of Mediterranean Europe Fernand Braudel, does not mean the economy of the entire world. That has effectively existed only since the eighteenth century at the earliest. Rather, it designates a large region which, through regular networks of exchange, has achieved a high level of economic integration and sustains a relatively autonomous division of labor internally. This relative autonomy enables a world-economy to constitute its own "world," self-sustaining and resilient in the face of alterations, but capable of linking to more distant zones as the value of the goods it circulates grows.[25]

This is how we can imagine the South China Sea world-economy: as a relatively autonomous but internally integrated trading zone that came into being in the second half of the fifteenth century, thanks to the organized penetration of Chinese merchants coming from the north and Muslim merchants coming from the south (Map 7). The Zheng He expeditions deserve some credit for enlarging Chinese participation in this zone,

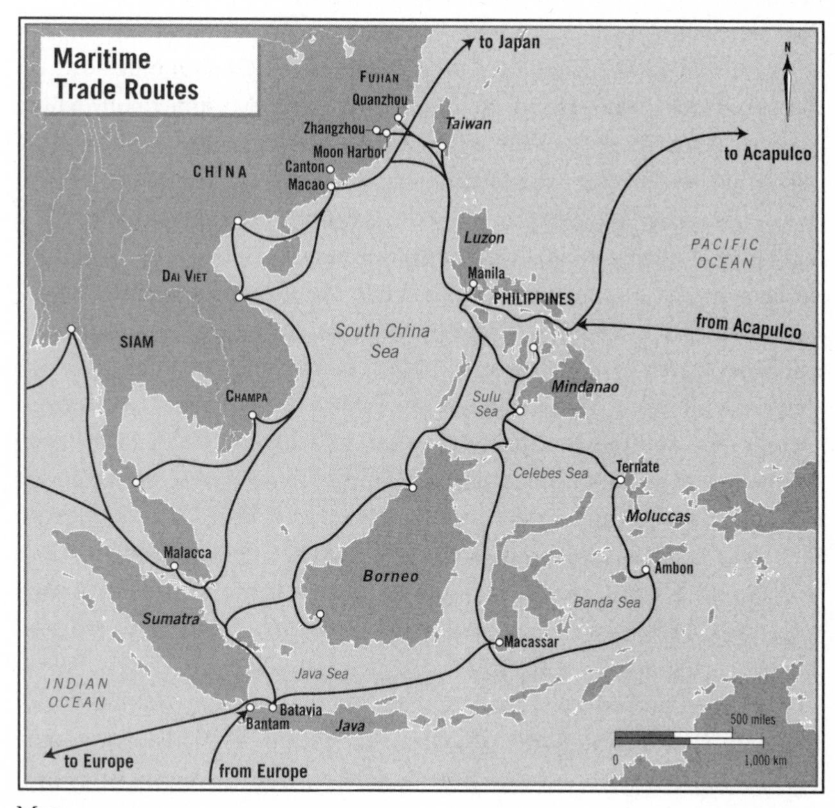

Map 7

but no amount of state voyaging could have created this world-economy. Trade had to surpass tribute for this to happen.

The trade was organized along two main routes, both starting at Moon Harbor and Quanzhou. The Eastern Sea (Dongyang) route headed for the lee of Taiwan; there one spur ran north to Japan, but the main flow of trade went south to the Philippines, down through the Moluccas (the Spice Islands to the Europeans) and west to Java. The Western Sea (Xiyang) route hugged the coast of the mainland past Vietnam, across the Gulf of Thailand, and on to Malacca. When Moon Harbor native Zhang Xie (1574–1640) compiled his survey of maritime trade in the 1610s, he structured the book around the two routes, which is why it is called *Study of the Eastern and Western Seas (Dongxi yang kao)*.[26] Zhang mentions rutters as one of the types of written sources he has consulted; no surprise that the Laud rutter organizes its data in exactly the same way.

The world-economy preceded the arrival of Europeans, which is why they were able to take advantage of the regional trade once they came on the scene. The Portuguese, sailing in from the west, finally got their perch on the tiny peninsula of Macao in 1557. The Spanish, coming across the Pacific from the west coast of the Americas, discovered the perfect harbor at Manila in 1570. They also discovered a trading community of over three hundred Chinese and the court of a minor Muslim rajah, whom they tricked and deposed the following year. The third major European player in this economy, the Dutch, reached the South China Sea only in the 1590s. After returning in the new century under the banner of the VOC, the Dutch East India Company, they set up their base of operations on Java, first at Bantam on the west end of the island in 1609, then at Jakarta (which they called Batavia) further east. Java gave a strategic position from which to lock down the Moluccas (the Spice Islands), but it left them too far from China, though not for want of trying. Their longest toehold was on Taiwan, where they set up a base in 1623 in order to compete with the Spanish colony in Manila. Chinese were drawn to the island as farmers and hunters once the Dutch established their base, with the ironic result, as historian Tonio Andrade has pointed out, that Taiwan would become "Chinese" only as the result of its colonization by the Dutch.[27] Eventually the Dutch were ousted by the maritime warlord Zheng Chenggong (Koxinga) in 1662.

The Dutch had two resources to maintain their presence in this world-economy. One was violence; this was what gained the VOC a monopoly over the hyperprofits of the spice trade. The other was a deftness in operating an extensive intra-regional trade, such that the company moved more goods between sites within the South China Sea and the Indian Ocean than they did between Asia and Europe. The business was profitable so long as Jakarta could monopolize its regional markets. But monopolies are time-limited, as rules shift and competitors move in to contest them. The strengthening of Chinese commercial networks throughout the region meant that by the middle of the eighteenth century, Chinese merchants had a stronger grip on the trade than the Dutch or the Spanish. At the same time, the British were increasing their presence in the region, quite overshadowing the Dutch. Jakarta became peripheral to the main commodity flows in and out of the region, hanging on as a remnant of a time when gunpowder empires were still viable.[28]

Silver

The Spanish and Portuguese were quite as ready as the Dutch to fight their way into the region, but what got them into exchange networks and kept them there was a commodity over which they, the Spanish in particular, had near-monopoly control and what they thought was an endless supply. It was also the commodity that the Ming economy valued above all else as the medium of exchange: silver. It came from mines in the Spanish possessions in the Americas, principally Potosí (in modern Bolivia) and Mexico. The level of production from these mines was extraordinary, especially from the 1580s when a new refinement process using mercury increased the yield of silver ore, and into the 1630s, when the more accessible deposits were becoming exhausted and production slipped. During these decades, Spain controlled silver in volumes large enough to fund their empire as well as to buy their way into the South China Sea economy. Within a few years of setting themselves up in Manila, the Spanish were bringing silver down from the Andes to the coast of Peru, shipping it up to Acapulco, and stowing it on board the one galleon that made the Pacific crossing at the end of every winter. Roughly three tons of silver crossed the Pacific on the Manila galleon annually in the 1580s. By the 1620s the annual flow had risen to twenty tons, thereafter falling to about ten tons.

Fujian merchants responded with alacrity, loading as much merchandise as they could warehouse onto junks and sailing it out to Manila to exchange for the precious metal. The annual departure of the cargo junks was timed to coincide with the spring arrival of the Manila galleon. After the ships had arrived on both sides, prices were negotiated, duties paid, and then the goods and silver switched holds. Both sides made sure to put to sea before the June monsoons created their annual havoc with ocean shipping. The bridge that connected Moon Harbor to Manila, Fujian to Peru, Ming to Spain, and China to Europe was made of silver.

The volume of silver that flowed out of Manila led to the rumor that the Spaniards had a mountain of silver in the Philippines. The imperial household eunuch Gao Cai, whom the emperor posted to Fujian to tax the overseas trade for his personal benefit, sent a mission in 1603 to investigate the truth of the rumor. He used the term everyone did, *jinshan*, silver mountain. *Jin* means "gold," but it was also the polite word for "silver," which is what Gao was looking for, not gold. The idea of a silver mountain at the edge of the South China Sea so obsessed the popular mind, even after its existence in the Philippines was disproven, that many Chinese destinations in the Americas and Australia earned the nickname Jinshan, conventionally mistranslated as Gold Mountain. San Francisco is still known in Chinese today as Old Gold Mountain. There was in fact a real silver mountain, but it rose above Potosí. Matteo Ricci marked it on the enormous world map he designed for his Chinese friends in 1602. He gave it its literal translation, Yinshan or Silver Mountain.

Silver was the perfect commodity from the European point of view. Its value when traded for gold was three times higher in China than at home, yielding arbitrage profits simply waiting to be plucked. In addition, the goods that the silver bought in Manila were acquired at a price far below what they sold for in Europe. The trade was also ideal from the Ming point of view, and for the roughly the same reasons, in reverse. The price differential was fantastic: a hundred catties of Huzhou silk in 1639 could be sold for a hundred ounces of silver in China but fetched two hundred from Spanish buyers in Manila.[29] And once the sale was completed, the costs of the transaction were over. The Chinese seller did not have to convert his pay into another currency or commodity. He could cash out his profits the moment the deal was closed.

These trading arrangements did not benefit everyone, of course. The investments necessary to work in this economy were so huge that the cost of failure became enormous. And when failures occurred, as they did regularly in a trade that depended on the happy conclusion of voyages across difficult oceans at vast distances, the effects could be catastrophic. Trade tensions in Manila in 1603 erupted into a full-scale battle between Spaniards and Chinese, which ended with the estimated deaths of twenty thousand Chinese. The scenario was repeated in 1639. The return galleon had sunk the previous year after leaving Manila, and the outbound galleon from Acapulco in 1639 also went down in a gale—losses that followed a year when the Ming government shut down the coast and forbade merchants from sailing abroad. The strain of insolvency weighed so heavily on both sides that when a group of Chinese farmers in the countryside revolted against their Spanish overlord, the entire region ignited in rebellion, resulting in casualties on the same scale as 1603.[30] Business recovered within a year or two, however. There was too much to be lost on both sides—all of it measured in silver—for a massacre to derail the trade.

How did all this silver affect the Ming? Even before the Spanish silver arrived, the Ming economy was already undergoing a commercial explosion that meant prosperity for many and envy for the rest. Attributing the explosion of wealth to the arrival of all this South American silver reverses cause and effect. It was the prosperity of that economy that attracted European buyers in the first place and persuaded them to surrender much of their precious metal in order to acquire Ming goods. On the other hand, the volume of silver coming from Manila and Macao as well as from Japan, where it was being produced for a time in almost equal volume, was so great that the Ming by late in the Wanli era was literally awash in money. As this commercial wealth outstripped other sources of income, merchant families were able to surpass the gentry in conspicuous consumption, if not exactly in cultural attainment. The old fourfold status ranking that put the gentry on top and the merchants at the bottom was being inverted. Silver may have been regarded as a tasteless acquisition in polite circles, but everyone wanted to acquire it.

The last decade of the Wanli era, the 1610s, was when anxiety about the spendthrift habits and atrocious taste of the *nouveaux riches* reached a peak. It was also, not coincidentally, the time when gentry authors were instructing the newly rich in the cultural habits they were expected to master if they hoped to enter polite society. The manual for tasteful consumption that Wen Zhenheng completed at the end of that decade, *Treatise on Superfluous Things,* is full of warnings about silver badly spent. Wen knew whereof he wrote, being the great-grandson of the great Wen Zhengming. His guidebook is driven by the certainty that uninformed consumers can go badly wrong when spending their wealth, and stresses how necessary it is to stick to his rules if you did not want your wealth to make you appear a complete boor.

Take, for example, Wen's instructions on how to conduct yourself while hosting a gathering at your private teahouse. The example is apt, for only the wealthiest could imagine owning enough urban real estate to lay out a garden large enough to build a teahouse in.[31] Wen's first piece of advice deals with the servants. "Train a boy to the exclusive service of tea," he advises, otherwise you will end up fussing with the tea things and getting distracted from your main task, which is "to spend the whole day there in pure talk, and the chilly night in sitting there in a dignified attitude." An additional note stresses that the evening pose "cannot be dispensed with." Alas, some people could only be expected to act badly, spilling the tea and lounging indecorously. Wen is strict on so many other subjects—parrots, for example. They "must be taught short poems and harmonious phrases," which means taking care not to allow them into such low-class spots as markets, wells, and villages, since the chattering that goes on there is "a violent assault upon the ear." Furniture was also treacherous. Stay away from dragons, he advised. Table legs carved into dragons were the height of vulgarity.[32]

Style was not the only thing that escaped convention as the tide of silver washed into Wanli society. As new money eroded the old certainties about social status, new ideas about how life could be lived were espoused. A friend of Zhang Xie, the chronicler of the sea routes out of Moon Harbor, expresses this new consciousness in a preface he wrote for Zhang's book. The mariners of Moon Harbor, he writes, "look upon the huge waves under the open sky as though they were standing on a steady hill, and gaze upon the sights of strange regions as though they were taking a stroll outside their own homes." They are "at ease on the ocean's waves, sailing their boats as though they were plowing fields." They address foreign potentates "as though talking to the neighbors."[33] The hoary classical trope about men plowing and women weaving no longer applied to the people of Moon Harbor. These were not the lives they lived.

Zhang did not expect that knowledge of the maritime world he documents in his *Study of the Eastern and Western Seas* would alter everyone's perceptions, but he does make one strong statement in his preface which hints that it should. Discussing the challenges of anthologizing a wide range of material as he has done, he singles out authors who simply requote old texts without paying any heed to recent developments and accuses them of perpetuating ignorance rather than creating knowledge. He wants to create knowledge, which is why he interviewed sailors to find out everything he could about trade routes around the South China

Sea. The book was not much taken up by readers at the time and had little impact on the knowledge that most people of the Ming thought important. The Laud rutter and the Selden map suffered the same neglect, which may go some way toward explaining why the sole copies survive in a library up the Thames.

Europeans in China

The flow of silver into the South China Sea world-economy brought with it a flow of strange people, from opulently dressed Portuguese with their African slaves and pet monkeys to a randomly assembled proletariat of sailors, soldiers, and smiths drawn or dragged into the irrationally dangerous business of global travel from all over the globe. The people of the Ming were fascinated. "The irises of their eyes are a deep green, and their bodies as white as freshly cut lard," Shen Defu writes of the Portuguese ("Franks") in his 1606 commonplace book, *Unofficial Gleanings of the Wanli Era*. "Of all the armed men of the seas, they are in general the most clever at gaining wealth, and not entirely by plunder either." When the Dutch arrived, even the people of the Ming were unprepared for how outlandish they looked. "Their appearance and clothing were unlike those of the earlier Islanders," by which Shen meant those bringing tribute from the islands of the Eastern and Western Seas, here signifying the Portuguese. "Because their beards are completely red, they are called the Red Hair foreigners."[34]

What grabbed Shen's attention more than the color of Dutch beards was the accuracy of their cannon. He notes that Ming sailors were caught off guard the first time they encountered a Dutch ship in coastal waters, which he dates to 1601. "They were unaware of their technical capacities, and so just went ahead and fired on them with the cannon they ordinarily used." The Dutch responded in kind, with stunning accuracy and to chilling effect. "They saw only a thread of greenish smoke, and then in an instant were reduced to a pulp." Shen allows that the Dutch had reason to open fire to protect their cargo, but suggests that such technology moved naval engagements to a whole new level. The Dutch "didn't so much as fire one arrow, yet the dead among the sailors were innumerable. And so they spread terror across the sea."[35] This is why Xu Guangqi and others argued strongly that the Ming should hire European gunners to improve its defense of the northern border.

The silver brought other Europeans to the Ming realm, not just green-

eyed merchants and red-haired cannoneers but Jesuit priests. These members of the Society of Jesus—a militant elite Catholic organization at spiritual war with all that the Protestant Reformation stood for—surfed the tide of global trade, intent on introducing Christianity wherever it took them. Their mission was a product of the globalizing economy, in two senses. First of all, it would have been unthinkable had Europeans not been engaging in maritime trade, thereby providing missionaries with routes to travel, ships to sail in, and ports to house mission bases. The Jesuits were the first to pursue this new opportunity with avid determination, sending the Spaniard Francis Xavier (1506–1552) with Portuguese merchants into the South China Sea in 1549, the year the society was founded. As the historian Liam Brockey has noted, the first turning point for the mission came in 1557 with the acquisition of a commercial toehold on Ming territory. Macao "was more than beneficial for the China mission," he observes. "It was of crucial importance for all missions of the Society of Jesus in East Asia." Wherever Portuguese merchants went, missionaries went with them, whether up the Pearl River to Guangzhou or across the East China Sea to Japan. Mission did not just follow trade but benefited from it.[36]

The Jesuit mission to Asia was also the product of the globalizing economy in terms of its financial operations. Bringing Christianity to unbelievers, the Jesuits understood, was an expensive operation: there were priests to educate, transport, and feed; residences, churches, and colleges to be built; supplies to be purchased and shipped; gifts to be given. The king and wealthy merchants of Portugal counted themselves among the patrons of the Jesuit mission, diverting a small modicum of their profits from the maritime trade to do so. But it would be a mistake to view the Jesuits as the passive beneficiaries of Portuguese trade into the South China Sea. They were active participants in elaborate currency arbitrage and commodity trading to support their ventures. A papal decree forbidding religious orders from engaging in commodity trading, intended to insulate missions from the losses that speculative ventures inevitably suffered, did not come down until 1669.[37]

Two Italians, Michele Ruggieri (1543–1607) and Matteo Ricci (1552–1610), were the first Jesuits to infiltrate themselves into the Ming realm. They managed to get permission to set up a church on the mistaken understanding of a regional official that they were some sort of Indian Buddhists. The building of the China mission would be a slow process that involved much inter-cultural negotiation and many false starts. For example, when cross-dressing as Buddhist monks seemed not to put the Jesuits in touch with the people they hoped to attract, they gave that up in favor of impersonating Confucians, very much to the benefit of their mission to the gentry. Finally, in 1601, Ricci would achieve his goal of setting up a mission church in Beijing.

The Macao connection was more than fortuitous for the Jesuits; it was essential. It gave them a base outside the Ming but close enough to be able to operate on the inside. Macao also provided access to the financial operations of Portuguese and Spanish trade flowing through the port. Moving wholly into China would have made that next to impossible. Suspicious of these foreigners, the Chinese read the Macao connection differently. They saw the port as the Jesuits' Achilles heel, their point of vulnerability. What could the connection possibly indicate except that they were in the service of the Portuguese, whose interests were not entirely commercial but political? As one aggressive official in the Ministry of Rites phrased this suspicion in 1616, "Their religion makes Macao its nest." It was widely believed that the Portuguese were bent on encroaching on the Ming realm, which meant that every Jesuit was a "cat's paw of the Franks."[38] Macao may have been an essential asset for the Jesuit mission, but in Chinese eyes it was a liability. Such was the contradiction at the mission's heart: it did not take place in an economic and political void but followed closely the contours of the economic and political power that made it possible.

Despite the hostility of powerful officials, many well-placed intellectuals in the late-Wanli generation interacted keenly with the Jesuits, some even converting to Christianity. Their motives were as varied as their personalities. As we have seen, some prized the knowledge that the Jesuits brought from Europe: geometry, astronomy, cartography, ballistics, hydrology—sciences of spatial calculation in which Europeans excelled. Some were intrigued by Christian cosmology, which interpreted Heavenly signs in a satisfyingly comprehensive fashion. Some admired the Jesuits' personal intellectual capacity and moral certainty, regarding them as fellow-travelers in the great program of improving the world.[39] The Jesuits had the good fortune to inaugurate their mission at a time when Ming intellectuals were struggling with fundamental questions about their own moral mission as well as basic technical problems of how to help the people through the two Wanli sloughs and how to defend the northern border against the forces that would bring the dynasty down in 1644. These were questions to which the highly educated Europeans appeared to have good answers.

The Society of Jesus was also fortunate in the man who ended up leading the mission. Matteo Ricci was subtle in his grasp of cultural patterns as well as strategic in his assessment of what a European in China had to do to achieve anything.[40] For example, he told Shen Defu, who lived near him in Beijing, that he had come to the capital "to present tribute."[41] This was not strictly true, Portugal not being a tributary state and Ricci not being Portuguese, but the statement was rhetorically effective by virtue of finding the right idiom in which to make his presence and ideas sensible. Ricci's effort, like Xavier's, ended in an immense failure, in Ricci's case the failure to gain an audience with the Wanli emperor. But it also produced the great achievement of devising a path that would enable Europeans to accommodate to Chinese values, and vice versa. Some other Catholic missionaries, particularly the Dominicans, were less tolerant of the culture into which they entered: less willing to find analogues for Christian habits that they mistook as fundamental truths, and ultimately less successful in persuading Ming intellectuals to trade their values and beliefs for an entirely different set. The Dominicans made considerable inroads among the people, though they survived only so long as the Christian communities they founded stayed beneath the radar of a state ever anxious that religion might be a smokescreen for sedition.[42]

The Fugitive's Return

The Suzhou constables sent south to find Guan Fangzhou were just about to wind up their manhunt for the silversmith when they decided to go out and have a final look around Macao. This was still in the 1570s, when the Portuguese had yet to construct the impressive fortifications that several decades later would convince some Chinese the foreigners were not to be trusted. Nor had Ricci yet arrived to start learning Chinese, which he did in 1582. While in Macao, the constables heard that a European wreck had just floated into port. It was without mast or rudder and appeared to be deserted. Curious, the constables went on board to have a look. They found two Chinese barely alive down in the powder magazine. By a remarkable coincidence that makes the story just a bit hard to believe, one of the men was none other Guan Fangzhou.

We never learn how Guan ended up on this European ship: as a captive? as a trader? as a stowaway? However it happened, once he washed into Macao, Guan realized that he might be able to turn his new situation to his advantage. The Ming had never formally surrendered sovereignty over the port to the Portuguese, yet Guan claimed a sort of extraterritoriality *avant la lettre* and assured the constables that they had no jurisdiction in Macao. The constables accepted if not the plausibility of the claim, then the reality that they were in no position to clap Guan in irons and carry him bodily out under the noses of the Portuguese. They would need to proceed by a more circuitous route to make an arrest. They had to convince Guan, not force him, to return. So they made up a story.

"We too were planning to go off and trade with the Europeans," they tell Guan. "But looking at you now, we see just how dangerous this is, so we've decided to go home. You can come with us if you like." The fugitive from justice hesitated: "What about the case against me?" "Your case has already been dismissed," they assured him. "It was covered by a general amnesty." Amnesties were a common practice when an emperor needed to ask Heaven for a favor—to relieve a drought, for example—or when the Ministry of Justice had too large a backlog to clear its cases through regular procedures. "You have nothing to worry about."

Guan's mistake was to believe them. Only later did the penny drop, but by then he had lost his freedom. They got him back to Suzhou, and just in time. The edict for Warden Wang's execution had just arrived. Guan's return meant that the silversmith, not the warden, would pay the penalty for embezzlement. It was the talk of the town, incontestable proof, so everyone declared, that Heaven worked in mysterious ways to put matters right. If that was true, then the mystery now included global trade.

10

COLLAPSE

ONE COLD night in the winter of 1657–1658, fourteen years after the Ming dynasty had been overthrown, Huang Zongxi (1610–1695) was awoken by a noise on his bookshelf. He lit a candle on his bedside table and peered at the bookshelf in time to see a rat scurry down from the shelf. He carried the light over to inspect what the rat had been chewing, and the guttering flame showed him that the rodent had selected a stack of *Capital Gazettes* from the Hongguang era (1644–1645). The capital gazette was the official newssheet that every court printed and distributed to its higher officials with information about events, policies, and appointments. The copies on Huang's bookshelf had been printed in Nanjing, where Hongguang for a time held his court. Hongguang was Zhu Yousong (1607–1646), formerly the Prince of Fu and cousin of the Chongzhen emperor, who committed suicide in 1644 to avoid being captured by rebels. The remnants of the Ming court fled south to Nanjing, where the prince was enthroned as the Hongguang emperor. The armies serving the invading Manchus descended on Nanjing and chased Hongguang out a year later. The Hongguang edition of the *Capital Gazette* was suspended.

Huang Zongxi had somehow managed to preserve his copies through the unsettled years that followed. He recalled the winter's night on which he saved them from the rat, because that was the moment when he decided that he could no longer put off compiling a record of the short reign. "I was saving them to use as documentation for writing the history of that period," Huang writes, "but in the years since that time I saw many troubles and was often ill." Now the danger was of losing the living memory of that first difficult year after the fall of the Ming. "What you can pick up now about the old times dwindles by the day," he worries. His own unsettled circumstances made matters worse. "I have moved three times in ten years, and many of the books I gathered have gone missing." If the book were not written soon, it never could be, for more time would pass and the memory of the reign would fade. "Who," he asks himself, "will take up this duty after I die?"[1]

Taking the Blame

By 1658, Huang Zongxi had established himself as the most important historian and constitutional theorist of his generation. It had been fourteen years since Chongzhen's suicide, followed by the Manchu takeover of Beijing. Many Ming officials bowed to circumstances and made the transition to a new master, but many did not, choosing instead to live out the rest of their lives as loyalists to the dynasty that had honored them with office before 1644. This was certainly the expectation in Huang's social and intellectual circle. Thirty-four at the time the Ming fell, he decided that the only way to remain loyal to the memory of the Ming was to decline to serve a second dynasty—just as a widow was not supposed to marry a second husband, though most widows did remarry and many officials did sign on with a second dynasty. Huang, however, withstood the pressure from conquerors and colleagues to offer his allegiance, and spent the rest of his life writing and teaching, leaving an intellectual and documentary legacy greater than anything he might have achieved through bureaucratic service.

Hongguang had not been a popular choice for emperor, as Huang notes in the *Veritable Record of the Hongguang Era* that he did end up writing. Shi Kefa (1601–1645), the tough-minded minister of war who struggled to rally the remnants of the Ming against the Manchu invaders and would die in the spectacular massacre of the city of Yangzhou in 1645, declared the Prince of Fu disqualified from emperorship on seven grounds: he was corrupt, licentious, alcoholic, unfilial, vicious to subordinates, unstudious, and meddlesome.[2] Shi's frank assessment was not enough to annul the prince's candidacy: the minister of war made the error of assuming that the head of an imperial autocracy had to be a person of merit. It was proximity to the founder's direct line of descent that mattered, not who might actually be personally qualified to hold the top position. What really troubled most courtiers about the Prince of Fu was not his character; it was his parentage. His father was Zhu Changxun (1586–1641), the prince at the center of the "foundation of the state" controversy whom Wanli had wanted to designate as the heir apparent to please the boy's mother, Lady Zheng. Lady Zheng was thus the Prince of Fu's grandmother. To officials who had held out against Wanli's choice during that constitutional struggle, it almost felt as though the old emperor was getting his revenge from beyond the grave.

The revenge was short-lived. The Hongguang emperor was betrayed to the Manchus after one brief year on the throne in Nanjing and died shortly later in confinement. Three other cousins were pushed forward to lead fugitive Ming regimes in the distant south. One of them even appealed to the Pope to send an army of deliverance to China, though the letter reached the Vatican long after the fate of the Ming was sealed. None of the resistance groups was able to withstand the onslaught of the armies the Manchus commanded—consisting in fact of surrendered Ming troops for the most part.

The new Qing regime was willing to show Chongzhen the posthumous respect he deserved as a legitimate emperor who, like a chaste widow, had chosen suicide over dishonor. His Manchu "successor," the Shunzhi emperor (r. 1644–1661), had a stele placed in front of Chongzhen's tomb that praised him for "having sacrificed his life on behalf of the nation" while those around him "lost their virtue and let their country perish."[3] The history of his reign was thus written in a way that blamed his advisors—the very men who disdained to switch sides and serve the conquerors. Dorgon (1612–1650), the commander of the Qing forces and uncle of the child-emperor of the new dynasty, summarized the politically correct version of events: "The Chongzhen emperor was all right. The trouble was that his military officers were of bogus merit and trumped up their victories, while his civil officials were greedy and broke the law. That is why he lost the empire." His suicide had been convenient in allowing them to claim the mandate of Heaven without having to exterminate the previous mandate-holders, so they built him a tomb alongside the tombs of his imperial ancestors, gave him all the posthumous honors due to an emperor, and declared his dynasty at an end.

Officially there would be no honors for his upstart cousin, and certainly no Veritable Record. If someone were to compile such a book, he would have to do it out of sight of the new regime. Indeed, the discovery that someone was compiling such a record could be construed as a challenge to Qing legitimacy, effectively an act of treason. This may have been a morally feeble reason to keep putting the job off, but it was nonetheless compelling. The failure to do so only further compounded the sense of humiliation and self-worthlessness that loyalists felt about their part in the dynasty's downfall. Suspecting that there had to have been something their generation could have done to keep the Ming afloat, they looked through their closets for signs of failure—and found any number of them: practical, attitudinal, intellectual, ethical. One Shanghai writer who was only a child when the Ming fell even attributed the collapse of the dynasty to a shift away from classical literary style. The dynasty was set on the road to disaster as early as the Wanli era, he declared. "Once literary style had become greatly corrupted, the fate of the nation followed in its wake."[4]

This explanation for the fall of the Ming makes for fine histrionics but poor history. Huang Zongxi was a good historian as well as a good loyalist, alert enough not to assume that the responsibility for the disaster was inescapably embedded in the habits and inclinations of his class. His view, contradicting Dorgon, was that the dynasty fell because a mediocre emperor had failed to take action against the eunuchs and incompetent bureaucrats who surrounded him. "When the emperor does not conform to the Way," Huang coldly observes a few lines later in the rat-in-the-night preface, "what can the riff-raff do but count the days toward their own destruction?" Nonetheless, Huang did not regard the failures at the Chongzhen court as the main story. These were no more than the circumstances of the fall. Beneath the mismanagement and moral slide lay the fundamental weaknesses of autocratic rule. Autocracy neglected the bond that should exist between ruler and people such that when disaster struck, neither could trust the other to find a way forward. That, in Huang's view, is what lay at the root of the Ming collapse.

This was not the sort of analysis through which most intellectuals were prepared to understand dynastic decline as they experienced it. More simply, they looked about them, dazed by the onslaught that arrived more swiftly than had the Mongols in an earlier time and trembling to think of what was to come. The poet Wang Wei (ca. 1600–ca. 1647) expressed the despair of her generation in these eight lines of parting she wrote for her husband as he left to join the resistance against the Manchus:

> Mist rises from the desolate grass;
> The moon descends into the cold stream.

The soul goes home as the autumn ends;
Sadness comes in the dim night.
When will this disquiet end?
Will my inner heart grow cold?
You point your oars to the edge of the sky;
Seeing you off, I feel hesitant and uncertain.[5]

Wang Wei pointed no fingers, simply described the conditions in which her family and class found themselves at the end of the dynasty. Such testimony has influenced modern historians, who have fashioned their accounts of the Ming around the tragedy of decline.[6] But it is worth asking whether the Ming was in decline before it fell. One can argue that it was, and we shall rehearse some of these arguments. Yet it is useful to distinguish the outcome from the conditions that produced it. Whether the Ming was in decline or not, it is difficult to imagine how matters could have turned out differently, given the circumstances. Some of the blame could go to the officials around the Chongzhen emperor for failing to take measures that might have stemmed the military and fiscal tide that turned against the regime. While our attention in this chapter will be on those who advised, schemed, and fought their way through the final decades of the dynasty, the conditions under which they acted shaped the courses of action they could follow. Indeed, the greatest puzzle might well be to figure out how the Ming remained standing for as long as it did.

The Wanli Sloughs

To tell this story, we need to go back to the reign of the Wanli emperor, enthroned in 1572 and dead in 1620. Contrary to the standard narrative of decline, the failings of the emperor may not be the place to start. There is evidence that the Wanli emperor was indecisive and politically inept, but we have reached the point when it is time to pull back from the little dramas at court and see the bigger picture. In the case of the Wanli era, the bigger picture involves two major downturns in the environment.

The first Wanli Slough of 1586–1588 was an environmental collapse on a scale that stunned the regime and established a new benchmark for social disaster. The regime was able to weather the catastrophe thanks to the reforms that Chief Grand Secretary Zhang Juzheng had imposed on the administration of the empire's finances at the start of the 1580s. By tracking unpaid taxes and blocking the promotion or transfer of magistrates who failed to clear back taxes in their counties, Zhang got the financial system operating as close to its peak of efficiency as the system allowed and left the Imperial Treasury well stocked with silver when he died in 1582.[7] These reserves helped the Wanli court weather the storm when disaster struck in 1587 and ride it out the following year. The shock of the slough remained a strong memory. When a large famine began to build in Henan province six years later, court and bureaucracy mounted a rapid response that relieved the shortfall before local distress could mushroom into a regional crisis.[8]

Two decades passed before the second Wanli Slough arrived in 1615. The two years preceding the slough were years of flood in north China; in the second of those years the weather turned cold. What initiated the slough was a confusing patchwork of severe drought in some places and severe flood in others. Petitions for relief started pouring into the central government from everywhere in the autumn of 1615. On November 25, two grand secretaries forwarded a summary of these reports to the Wanli emperor. "Although the situation differs in each place, all tell of localities gripped by disaster, the people in flight, brigands roaming at will, and the corpses of the famished littering the roads, and not one report does not plead to receive the favor of your imperial grace." The emperor agreed to forward their summary to the Ministry of Revenue, which came back with a recommendation that massive relief be undertaken.

Shandong was hit by the famine worse than any other province. A report that reached the court in February estimated that over 900,000 people were on the brink of starvation, that local relief supplies had run out, and that civil order had completely collapsed. In March 1616, a lower degree-holder in Shandong province submitted an *Illustrated Handbook of the Great Starvation of the People of Shandong* to the court. The court diary noted that each picture was captioned with a poem of lament. A couplet in one of these poems became the tagline for the entire disaster.[9]

> Mothers eat their children's corpses,
> Wives strip off their dead husbands' flesh.

The famine moved from north China to the Yangzi valley later that year, reached down to Guangdong the next, and gripped the northwest and southwest the year after that. The worst may have been over by 1618, yet drought and locusts continued to harry the realm through the last two years of the reign. To this litany of disasters may be added massive sandstorms in 1618 and 1619—offshoots of the deforestation of the northwest. The one that blew in over Beijing at dusk on April 5, 1618, was so powerful that, according to the *History of the Ming,* "it rained soil. The air was thick, as though with fog or mist, and the soil kept raining into the night." A year less a day later, "from noon until night, sandy dust filled the sky, coloring it a reddish-yellow."[10]

Wanli died in 1620, just at the moment when the long run of cold, dry years came to an end. The crown prince, the legitimate one, was enthroned as the Taichang emperor. Within a month, even before his father had been properly buried, Taichang was dead. Yet another constitutional crisis threw the court into chaos. The succession was simple enough, but the son who was put on the throne as the Tianqi emperor was immature and untutored. During the next seven years (1621–1627), the realm fell into the grip of his chief eunuch, Wei Zhongxian (1568–1627). The political climate was evil, but the weather was remarkably normal. The last two years of the Tianqi era were wetter than usual, but without serious flooding. Nature's only major aberrations were earthquakes, which rattled every year of the reign.

The chaotic reign of the Tianqi emperor ended with his early death in 1627, to the great relief of almost everyone at court. His failure to produce a son could have thrown the regime into yet another constitutional crisis, but he had a sixteen-year-old brother to succeed him, a young man whose enthronement as the Chongzhen emperor led many to hope that here at last was an autocrat with whom they could work. But conditions would worsen, and Chongzhen had little chance to reverse the fate of being the last emperor.

The Northern Border

The people of the Ming were not alone in facing the famines of the Wanli era. The drought that gripped north China in those years also parched Liaodong, the region northeast of the Great Wall subsequently known as Manchuria. It was there that the Jurchen leader Nurhaci (1559–1626) was gradually building an ever broader alliance of Jurchen and Mongol tribes into a confederacy that in 1636 would take the new ethnic name of Manchu. Nurhaci was still submitting tribute to the Ming as late as 1615, but he was doing so as a cover for his own territorial ambitions. The drought and cold may have been what convinced him to stop sending tribute. Rather than withdraw as a more timid leader might have done, he escalated his competition with the Ming for Liaodong. He needed the grain that was grown there and was prepared to fight the Ming for it. The turning point came in May 1618, when Nurhaci launched a surprise attack in eastern Liaodong that led to the death of the commander-in-chief of the Ming forces and gave the Jurchens control of the region.

The Ming launched a major campaign against Nurhaci the following spring, but it was beset with difficulties. It was underfunded, because the Wanli emperor refused to disburse imperial household funds at the level required. It was also bogged down by snow, an effect of the colder weather. Barely a month after it had started at the battle of Sarhu on April 14, 1619, the campaign collapsed. That fiasco was followed by what the great fiscal historian Ray Huang has characterized as "a series of dazzling victories in one battle after another" for Nurhaci's forces. It was the beginning of the eventual loss to the Ming of all its territories beyond the Great Wall, though that loss would take another two decades to run out. That dry summer, three months before his death, Wanli explained to a grand secretary that the cause of the defeat was discord between civilian and military officials in Liaodong. Ray Huang, on the other hand, placed the blame squarely on the emperor. Wanli's refusal to release silver from the Imperial Treasury obliged the Ministry of Revenue to impose a temporary surtax on the land tax to pay for the Liaodong campaign. Not only would that surtax not be rescinded; it would be increased, as the next quarter-century of military misadventures and environmental disasters heaped impossible demands on imperial finances.[11] The military disaster at Sarhu meant that the military threat would continue to escalate, and that whatever the Ming had spent on defense, it would now have to spend more.[12]

The military problem seemed easier to fix than the more complicated and intractable problem of finances, and more than one official stepped forward with proposals. One of them was Xu Guangqi, the Christian disciple of Matteo Ricci. Xu in 1619 began his determined campaign to argue that the most effective way to enhance the Ming's military capacity was to borrow the best European knowledge.[13] His program included not just ballistic technology but the science of Euclidean geometry, which would help gunners improve their sighting. Xu had earlier helped Ricci translate the first six books of Euclid's *Elements* into Chinese, which was published in 1608. He also advocated bringing Portuguese soldiers up from Macao to train Chinese gunners in the newest methods. When a

Portuguese gunner in 1622 scored a direct hit on the Dutch gunpowder store during a failed Dutch attack on Macao, word got out that it was the calculations of the Italian Jesuit Giacomo Rho that had been responsible for the hit, and this was all the proof Xu needed. He obtained permission for seven gunners plus an interpreter (actually the Jesuit missionary to Japan, João Rodrígues) and an entourage of sixteen men to travel up to Beijing that year.

The question of whether to rely on foreign technology—and perhaps the even more pressing worry that foreign soldiers in Beijing would gain knowledge they could later put to military use against the Ming—provoked a controversy at court that threatened to destabilize the entire project. When a cannon exploded during a demonstration the following year, killing the Portuguese gunner and wounding three Chinese assistants, the project was canceled and the gunners were sent back to Macao. The experiment was repeated six years later when Xu was able to gain permission to bring a second team of gunners, with Rodrígues along again as interpreter, to Beijing. Opponents at court succeeded in blocking the delegation when they got to Nanjing, but the Chongzhen emperor eventually issued the edict permitting them to proceed to the capital. Indeed, they should do so with haste, as Jurchen raiding parties were roaming around the capital region.

The edict reached them on February 14, 1630, and they set off. Sixty-five kilometers short of Beijing, outside the city of Zhuozhou, the gunners encountered one of these Jurchen raiding parties. The Portuguese contingent retreated inside the city gates, mounted eight of their cannon on the wall, and fired when the Jurchens rode into range. The show of artillery had its effect, and the raiders departed. It was enough to win over some of those at court who still doubted the wisdom of bringing foreigners inside the realm.[14] It also emboldened Xu to ask the emperor to send Rodrígues back to Macao for more gunners and more cannon, and to permit Giacomo Rho, the Italian mathematician who defeated the Dutch in 1622, to enter the capital and take up a post at the Astronomical Bureau.

Xu's project to engage foreigners, a politically delicate maneuver, was badly shaken when twelve of the Portuguese soldiers were killed in a military revolt in Shandong in 1632 and Xu's chief military disciple was executed for failing to suppress the revolt. The debacle let loose a firestorm of factional politics that had nothing to do with the military situation the regime faced and everything to do with one clique trying to destroy the other.[15] Xu's initiative on its own was not enough to shift the military balance in Liaodong. He was absolutely correct in realizing that firearms would be decisive in future battles, but without an emperor able to direct the defense of the realm, a grand secretary who enjoyed the confidence of his peers, or a military commander immune from impeachment for reversals in the campaign, technical knowledge would not change the tide of events.

The garrison command at Guangning fell to the Jurchens in 1622. Ming forces had to withdraw inside Shanhai Guan, the Gate of the Mountains and Seas, the eastern terminus of the Great Wall where it meets the sea. But colder, drier weather led to food shortages in Liaodong, obliging the Jurchens to pull back and rebuild. This retreat gave the Ming dynasty a chance to catch its breath and to cast about for ways to fund its border defense. An increase on current levies seemed untenable. As a capital official reported to the Tianqi emperor in the summer of 1623, "The costs of military supplies and courier deployment in Liaodong have escalated so greatly that the material strength of the entire realm goes into supporting this one small corner of it." Consequently, "the common people year after year have to scrape the marrow out of their bones and sell their children and wives to meet the harsh exactions."[16] The Chongzhen emperor attempted to address the problem by tightening the tax system and reducing abuses among the privileged. He also tried to ensure the flow of revenue to the center by blocking the careers of field administrators who did not deliver their quotas, though this order only had the effect of increasing the bribes that field administrators paid to the clerks in the Ministry of Revenue to hide their shortfalls.[17]

The Ming forces were able to take advantage of the Jurchen fall-back to recapture some of Liaodong. A swashbuckler named Mao Wenlong even succeeded in humiliating the Jurchens by invading their sacred homeland in the Ever White Mountains in 1624 (incidentally a habitat for Siberian tigers). Nurhaci's death in 1626 further stalled Jurchen expansion, and the Jurchens turned to other means, including diplomacy. They sent a letter to Mao, hoping to persuade him to switch sides. The letter begins by pointing out that disasters have always portended the fall of a state. The Ming, whom the letter refers to insultingly as the "southern dynasty," was experiencing its full share. "As the southern dynasty approaches its end, the number of deaths is endless and even the imperial emissaries die, so how can one general save the situation?" Then follows the invitation to switch sides. "The healthy animal finds a tree and climbs it; the wise minister finds a ruler and serves him." The letter concludes by observing that "the southern dynasty has lived to the end of its natural life; its time and course are exhausted. This is not even worth regretting."[18]

Mao did not reply, probably because he figured he was on the winning side. The following February, however, the Jurchens launched an offensive against Korea, forcing Mao to pull back. He may have ceded territory, but his new position at the mouth of the Yalu River placed him in control of the profitable maritime trade between Liaodong and Shandong, giving him the means to assert implicit autonomy as a semi-warlord. The Jurchens quietly reopened back channels to see whether he could be induced to come over to them. Mao was sufficiently well supported by rents from the maritime trade that he could afford to play each side against the other, and he did so until 1629, when his superior officer Yuan Chonghuan (1584–1630), suspicious of Mao's intentions, used the pretext of carrying out an official inspection to enter his camp and order one of his officers to behead him on the spot. "The murder of Mao Wenlong," the historian Frederic Wakeman has noted, "threw the frontier into turmoil, ultimately releasing many of the general's freebooters to plunder on their own."[19]

Yuan's dramatic act may have prevented Mao from switching sides, but the turmoil distracted him from detecting the swift offensive that Nurhaci's son Hong Taiji was preparing. That November he went around Yuan's defensive position and dispatched contingents of mounted archers onto the North China Plain. One contingent rode right to the walls of Beijing. Another attacked the city of Zhuozhou further south, where as we have noted Xu Guangqi's Portuguese gunners fired on them. The Jurchen raiding parties were not prepared to back up their invasion and withdrew beyond the Great Wall, but the court needed someone to blame. What better scapegoat than the man who murdered Mao? Yuan Chonghuan was recalled to Beijing and subjected to the humiliating punishment of beheading and dismemberment the following January. His crime was the traitorous act of failing to stop the Jurchens from reaching Beijing. It was a crime for which many other officers would pay with their lives in the coming years.[20]

Hong Taiji was able to launch his offensive on the strength of having devoted the three years after his father's death to reconsolidating the Jurchen forces under his leadership. Although he withdrew his forces at the end of that winter, he had demonstrated that the Ming military presence in Liaodong was ineffective. Gradually he exerted full control over the larger region of Manchuria. In 1636, he felt confident enough to found a dynasty, the Qing, and have himself declared emperor. The symbolism of the new dynastic name implied that the Qing, a water image meaning clear or pure, would submerge the Ming, a fire image of sun and moon together. Whether Hong Taiji believed that his dynasty would become more than the sort of regional regime the Jurchens had commanded in north China four centuries earlier we do not know, but the dynastic founding was at least a challenge to the Ming. Hong Taiji died in 1643 before he could carry out his conquest. The succession passed to his young son, and the campaign to his brother Dorgon.

The Chongzhen Slough

We move now from the actors on the Chongzhen stage to the stage itself, the environment. No emperor of the Yuan or Ming faced climatic conditions as abnormal and severe as Chongzhen had the misfortune of doing. In the first years of the reign, the difficulties were mostly confined to the northwest, especially the province of Shaanxi. Drought and famine were so severe that a censor reported at the end of 1628 that the entire province was a disaster zone. Temperatures plummeted the following year, as a cold spell set in that lasted into the 1640s. It was felt not only by the people of the Ming. During the 1630s, Russians experienced severe cold for at least one of the three months of December, January, and February. In the 1640s, however, severe cold was reported for every month of winter, making this the coldest decade in Russian history since the twelfth century.[21] Lying between China and Russia, Manchuria suffered the same fierce cold. The Jurchens may have been drawn south by the wealth of the Ming, but they were also pushed south by the cold.

The first serious famines began in 1632, the fifth year of the Chongzhen era. The court that year was inundated with memorial after memorial reporting on extraordinary conditions all over the country and the extreme social dislocation that went with it. "Banditry everywhere, and every day worse than the day before," exclaimed an official sent to inspect the northwest. "Communication between north and south has been almost completely cut off," reported an official assigned to deal with the disaster along the middle section of the Grand Canal. "The poor flee and become brigands while the rich slip off undetected," declared a second in the same region. "Merchants are not moving goods, and all the roads are blocked."[22]

After 1632, the disaster only deepened. Locusts began to appear on a massive scale in the eighth year, 1635. Then finally the dry weather turned to full-scale drought in the tenth year, 1637. For seven years running, the Ming suffered droughts on an unprecedented scale. During the great drought that devastated western Shandong in the summer of 1640, the famished stripped the bark off trees to have something to eat, then turned to rotting corpses.[23] In the commercial city of Linqing in northwestern Shandong, the desperate resorted to cannibalism.[24] Famine spread its pall southward over the Yangzi delta the next summer. A telegraphic entry in the Shanghai county gazetteer describes the scale of the disaster:

> Massive drought.
> Locusts.
> The price of millet soared.
> The corpses of the starved lay in the streets.
> Grain reached three-tenths to four-tenths of an ounce of silver per peck.[25]

The drought continued for another two years. Desperate to turn the tide, the Chongzhen emperor on June 24, 1643, issued an edict commanding all his subjects, from the highest official to the lowest day laborer, to purge the evil thoughts lurking in their hearts so that Heaven might be persuaded to end the punishment of drought and bring back the rain.[26]

Epidemics followed in the wake of drought and famine. Much of it was due to smallpox. Chinese were already managing the disease by practicing variolation, a simple form of inoculation, but the Jurchens/Manchus were not. They had a particular dread of this disease and were so anxious to avoid coming into contact with infected people that, at several key moments in their military incursions through the 1630s, they fell back from an area in which contagion had been reported. Fear of smallpox was partly what ended Hong Taiji's foray onto the North China Plain in 1629–1630.[27] The epidemic that scourged the region around the Gate of the Mountains and Seas in 1635 was probably smallpox. Smallpox broke out in Shandong in 1639 on a scale sufficient to convince the Manchus to cancel that winter's raid into north China.

Epidemics struck other regions of the country as well. The northwest was particularly hard hit. The first massive epidemic in that region devastated Shanxi province in 1633. Three years later it spread through

Shaanxi and southern Mongolia. In 1640, all Shaanxi was infected. After it was over, provincial officials estimated that 80 to 90 percent of the population died.[28] Though the percentage surely exaggerates the actual toll, it does indicate the severity of the episode, at least in some parts of the province. Whether the disease was plague is much debated. A strange explosion in the rat population in 1634 in the far northwest—the *History of the Ming* reports that a hundred thousand rats surged across the Ningxia countryside eating everything in sight—has encouraged some historians to connect the rats to the outbreak.[29] Whether the two events were connected, and whether the rats were carrying plague-infected fleas, are still anyone's guess.

A severe epidemic struck the Yangzi valley in 1639, and again an exodus of rats in the mid-Yangzi region the same year has raised the specter of plague. The sickness returned with even greater virulence two years later, not just in the Yangzi valley but throughout the eastern half of the realm. For one Shandong county that year it was reported that well over half the residents of the county died of the sickness. To the entry reporting the epidemic, the compiler of the local gazetteer has added this desperate note: "Among all the strange occurrences of disaster and rebellion, there had never before been anything worse than this."[30] In another Shandong county south of the Yellow River, where the epidemic completely exterminated some villages, an estimated 70 percent of the people died; the same percentage was recorded as well further up the Yellow River valley in Henan.[31] Locusts at the end of the summer then cleared the land of every edible plant, leaving absolutely nothing to eat.

The epidemic seems to have paused briefly in 1642, then resumed annually, devastating communities all the way from Jiangnan in the south to the border in the north.[32] It was understood at the time that Beijing was the epicenter of these waves of sickness, and that the Grand Canal, once the great avenue of national prosperity, was now the highway for the infected to spread the disease from the north. The effect of the epidemic on top of famine was deadly. "The great majority of the people have died" is a phrase much repeated in local records of these last years. "Of every ten homes, nine are empty" is another. As 1644 dawned, every county in northern Shanxi was infected.[33]

This was the Chongzhen Slough, the most prolonged series of disasters since the Taiding Slough in the 1320s. Crops withered, food supplies dwindled, and the commercial economy shut down, driving the price of grain to unprecedented levels. People had nothing with which to pay their taxes. A hardship for them, it was worse for the government, which found itself without the means to pay the soldiers who defended the border or the courier soldiers who kept the machinery of state moving. As early as 1623, the minister of war informed the emperor that the courier system was completely exhausted. Stringent new rules about who had the right to use the system needed to be applied if state communications were not to break down altogether.[34] But this was not sufficient to ease the burden, and so the ministry took the radical step in 1629 of closing some courier stations to save the cost of staffing them. Realistically, no amount of tightening was going to meet the unrelenting costs of waging war in Manchuria. The state saw no alternative to levying heavier and heavier taxes to keep pace with the soaring military costs. Black humor punned on the word Chongzhen/*chongzheng* ("double levy") and called it the Double Taxation era.[35] When 1644 arrived, 80 percent of counties had stopped forwarding any taxes at all. The central treasury was empty.

Rebellion

The financial meltdown hit hardest the northern areas that depended on central allocations to keep operating. At the beginning of the Chongzhen era, they were the first to suffer famine. Belt-tightening left soldiers and couriers without pay or rations. Many simply abandoned their posts, fleeing to peripheral regions where they could survive between day laboring and banditry without being tracked down. When drought struck one of these peripheral regions, Shaanxi province, in the spring of 1628, some of these men mutinied. This was the beginning of a tide of rebellion that would wash back and forth over the realm for the next seventeen years.[36]

With every mutiny and every successful raid on a government granary or a county yamen, the men who turned their back on the Ming and took survival into their own hands gained the confidence to go on to more ambitious conquests. Two rebel leaders came to command large followings and eventually declare their own short-lived dynasties, Li Zicheng (1605–1645) and Zhang Xianzhong (1605–1647). Li and Zhang were both from small communities in the impoverished north of drought-prone Shaanxi. Li got a job at a postal station in 1627 but lost it when the station was closed two years later. He worked as a tax collector for a time, flirted with soldiering, then drifted into banditry. Zhang's early years have spawned more dramatic stories. His pock-marked face may signify that he suffered smallpox as a child and survived. While a teenager, he was disowned by his family and thrown out of his community, according to one story, after killing a classmate. The story may be apocryphal, but the part about going to school seems to have been true, for two Jesuit missionaries who met Zhang near the end of his inglorious career discovered he was literate. As the safest place for a violent young man was in the army, Zhang became a soldier. He was accused, possibly unjustly, of plotting to mutiny against his commanding officer. Another officer intervened and saved him from execution, so the story goes, but he was booted out of the service. Having no skills other than fighting, in the summer of 1630 he turned to the only other career open to a man of his talents, banditry.

Li and Zhang were among the many marginal young men who formed and re-formed bandit gangs over the next few years across north China. Gradually these gangs linked into loose armies and, as they did, sought territorial bases from which to draw revenue and defend themselves against the armies the Ming sent to suppress them. In the end, none of the aspiring peasant warlords was successful in establishing a permanent regime. Even those who set up civil administrations remained in the end peripatetic, sometimes moving as new opportunities arose, sometimes picking up and fleeing as the forces sent to quell them moved in. By the mid-1630s, these northern armies probed down through Henan and Anhui into the Yangzi valley. Both Li Zicheng and Zhang Xianzhong suffered major defeats in 1638 at the hands of Ming armies. Were it not for the many burdens the state faced at this juncture, neither should have been able to revive their war machines.

But they did. Within two years, Li and Zhang had rebuilt their mobile regimes to a level that allowed both to harbor dynastic ambitions. Neither, however, was able to assert unchallenged control over any particular territory. Both moved around the interior of north China, from Henan to Shaanxi and down to Huguang, depending on the movement of Ming armies. As 1644 dawned, Zhang was down in Huguang after a failed attempt to take Nanjing and was preparing to move west into the inland fortress of Sichuan. Li, however, had just captured the ancient capital city of Xi'an. There he inaugurated the Shun (Submission) dynasty—though whether the submission was of Li to Heaven or of Heaven to Li was a matter of opinion—then late that winter launched a full-scale invasion of Shanxi province. While there, he looked even farther east, toward an undefended Beijing, and decided to make a bold and unexpected dash on the capital.[37]

The Chongzhen emperor issued a general mobilization order to all military commands on April 5, but the response was too weak to defend the capital city. Beijing fell to Li's forces on April 24. The emperor and his family retreated to the inner recesses of the Forbidden City. Seeing no way out, Chongzhen slew his daughter and retreated to Coal Hill behind the Forbidden City, where he hanged himself from a tree. The news of his death shocked the realm. The lunar date—the nineteenth day of the third month—burned itself into public memory. Chongzhen's suicide could not be safely mourned under the next dynasty. It had to be sublimated into some other commemoration, and was. Within a few years, there sprang up throughout Jiangnan a cult that worshipped the sun as it rose every year on this day.[38]

The news of Li Zicheng's attack on Beijing reached the Gate of the Mountains and Seas, where the commander of the Ming border forces, Wu Sangui (1612–1678), was holding back the Manchus. Wu decided on a desperate course. He approached Dorgon, the Manchu commander on the other side of the gate, with a proposal. The two generals would suspend hostilities, and for great honors and a mammoth reward, Dorgon would join him in a massive counterattack to drive the rebels from the capital. For Wu, this was a provisional arrangement to meet an unexpected crisis; for Dorgon, it was the nail that sealed the Ming coffin. Facing the prospect of fighting a joint force far greater than his own, Li installed himself as emperor on June 3 at the last possible moment, then beat a hasty retreat the following day. The day after that, the Manchus entered the Forbidden City and declared the inauguration of the Qing dynasty. Li Zicheng died a year later on the run.

Meanwhile, Zhang Xianzhong retreated into Sichuan, where he founded the Great Kingdom of the West, a notorious regime that operated a reign of terror for two years in order to support itself. Zhang's sole concession to humanity, as later reports would have it, was to permit the two Jesuit missionaries he found in Sichuan when he arrived to baptize as many people as they liked before he executed them en masse. In November 1646, he was forced by the Manchus to abandon the province and flee north. Manchu soldiers caught up with him and killed him on January 2, 1647.

The fall of the Ming dynasty is many histories: the history of the expansion of the Manchu empire on the northeast border, the history of the most massive rebellions to wash over China since the fourteenth century, the history of the disintegration of the Ming state, and the history of a major climate episode. Different in the stories they tell, they overlap and together constitute the same history. Could the rebel armies of Li Zicheng have taken control of the Yellow River valley in 1641 had an epidemic not wiped out 70 percent of the population earlier that year, leaving the region undefended, for example?[39] To decide which destroyed the dynasty—fiscal insolvency? rebellions? Manchu military might? the weather?—would exclude more truths than it would illuminate. At this particular conjuncture in China's past, their combination was what brought down the house of the Ming. Perhaps the greater puzzle than deciding which event destroyed the dynasty is asking how the Ming managed to survive as long as it did.

Living through the End

The people of the Ming came out the other side of the Chongzhen Slough to find themselves the subjects of a new dynasty. The routes to the world beyond 1644 were many, some easier than others. The vast majority accepted their fate, submitted to Manchu authority, and, if they were men, displayed their personal submission to the new dynasty by adopting the nomad hairstyle of shaving the front of the head and growing a queue at the back. It was a humiliation, but when Dorgon in 1645 declared it to be the price of keeping one's head, few resisted. Some did, however, keeping alive for several years the hope that the Ming dynasty might be restored.

As the Zhu family had nothing to gain by submitting to the Manchus, except perhaps their lives, some of the princes lent their persons to the resistance. The crown prince fled Beijing late in 1643 but was captured by rebels. When the dynasty tried to find a cousin to succeed Chongzhen and continue the line, only two were deemed suitable and available. In-fighting among court factions determined that the Prince of Fu would become the Hongguang emperor. He lasted on the throne for a year, but his armies could not hold back the Manchus, and they captured him outside Nanjing. The succession was passed up to a great-uncle in a distant collateral line (the Longwu emperor, r. 1645–1646), who also lasted only a year, then slid sideways to his brother (the Shaowu emperor, r. 1646), and then back down to a cousin of the Prince of Fu (the Yongli emperor, r. 1646–1662). These were the emperors of the tail to the Ming dynasty known as the Southern Ming.[40]

The last pretender, Yongli, was forced to flee into Burma in 1659 to escape the armies of none other than Wu Sangui, the general who invited the Manchus through the Gate of the Mountains and Seas in 1644. Wu was still in their service, though he would rebel in 1673 when the second Qing emperor decided to shut down the large fiefs that had been given to the Chinese military leaders who had put his father on the throne. Even in Burma, Yongli was apprehended. He and his teenage son were taken under armed guard back to Beijing, but on the way, in May 1662, it was decided that they be executed for fear that their presence back in the country would inflame anti-Qing resistance. After that, no other Zhu male dared look for a throne.

During the first year after the fall of Beijing, there was hope that military resistance might turn the tide against the Manchus. There was no effective coordination of these efforts, however, so that one city after another fell to the forces of the new Qing dynasty as they pressed southward to the Yangzi and beyond. The momentum of this invasion, unlike the Mongol invasion four centuries earlier, became unstoppable. The Manchus announced that cities surrendering without a fight would be leniently treated, and those resisting would have their citizens massacred. Many local leaders, seeing no way out, chose to capitulate peacefully. A few did not, and the Manchus were as good as their word. The first spectacular slaughter took place in the city of Yangzhou at the south end of the Grand Canal just above its junction with the Yangzi River. The second was across the river in the city of Jiading. Nanjing submitted without a fight, which allowed Qing forces to continue up the Yangzi River and then south through Jiangxi province. The last major resistance the Qing met in this region was at the provincial capital of Nanchang, which came under siege in the summer of 1645. Food supplies dwindled, and so soldiers inside the city were sent out on charges against the Manchus, but every time they were ineffective in breaking the siege. The men organizing the defense then turned to an itinerant monk calling himself Mahaprajna, who claimed he could defeat the Manchus by sending a dozen boys out onto the battlefield carrying long sticks of incense and reciting the *Prajnaparamita Sutra*. As the Manchus were devils rather than humans, the force of the boys' purity would dispel them, he claimed. The tactic was tried, alas, and the boys were slain below the city walls. When the provincial capital finally fell in February 1646, hundreds of thousands were butchered in reprisal for resisting the Great Qing.[41]

As the invasion advanced, the resistance had to withdraw further south and then southwest to evade annihilation by the Manchus. Their struggles have left a wealth of stories of heroic bravery and tragic defeat, all of them ending invariably in execution or suicide.[42] The crisis point in many of these stories comes when the demand is made to cut one's hair in the Qing fashion. A resistance fighter who retreated to the coastal islands of the Zhoushan Archipelago south of Shanghai wrote this poem before committing suicide in October 1644:

> Keeping one's hair divides Huns from Chinese;
> Supporting the Ming makes death and life one.
> Being the last loyal subject is my only achievement;
> Righteousness is all, this body nothing.[43]

Seven years later, Zhoushan served as the base of a second wave of resistance, but that attempt also went down in defeat. One of the men involved in this resistance drew the same ethnic line in the sand over the issue of hair. The Qing commander who captured him offered to spare him if he would cut his hair and submit to the Qing. "If I could have cut my hair earlier," he retorted, "why would I have waited until today?" For offending the ruling dynasty, the commander ordered his soldiers to cut off the man's arms and legs and leave him to die.[44]

A quieter mode of resistance against the order to adopt the Manchu hair style was to shave all one's hair off, effectively taking the tonsure of a Buddhist monk. This act was accepted as a sign of undertaking a religious life, and many chose this course of passive resistance. Most were what we might call political monks and did not take religious vows. The new regime could not begin to round up every monk and determine whether he was a man of faith or a man of resistance. Extracting the political monks from the real monks would have caused enormous trouble and further unrest, so the Manchus wisely decided to let them be and leave this one option for refusal open. Some men followed this course well after the Ming was gone. Shitao Daoji (1642–1708) was a member of the imperial Zhu family. He was barely two years old when the dynasty fell, but spent his formative years fleeing the Manchus in southwest China. He ended up becoming a political monk, but also a painter, arguably the most creative artist of the early Qing.[45]

Most people did nothing of this sort, of course. They had lives to get on with and obligations to meet. By 1646, after the collapse of two legitimate Southern Ming courts and many other illegitimate bids, most regarded the continuing defense of the Ming as a futile cause. Madam Huang Yuanjie, who is counted among the great poets of the mid-seventeenth century, composed a poem on Qingming Festival, a festival in the lunar calendar that fell that year on April 4. Qingming was the day when families gathered at their ancestors' graves to tidy them and to eat a meal of cold food to remember the hardships the dead had suffered. By 1646, everyone on the Yangzi delta, and in many other places in the country, had relatives and friends to recall who had died in the fall of the Ming. Huang had lost contact with her impoverished husband in the turmoil of the Manchu occupation of the delta the year before, never to find him again. Remembering him that Qingming day, she also chastised those too eager to forget:

> Leaning against a pillar, I am besieged with worries about the nation;
> Others, as always, go to the pleasure houses.
> My thoughts persist like unending drizzle;
> Tears fall like fluttering petals without end.
> Since we parted, a new year has already arrived;
> We still observe the custom of not lighting the fire.
> Thinking of my family I stare off into the white clouds,
> My small heart overwhelmed by grief.[46]

The same turmoil in 1645 claimed the husband of her bosom friend and sometime patron Shang Jinglan (1604–ca. 1680), an eminent poet in her own right. Shang's husband, Qi Biaojia (1602–1647), is the better known of the couple. Qi had been a prominent statecraft activist and local philanthropist dedicated to improving the age in which he found himself, and had died when the armies of the Qing overran his home county. Shang's poem of remembrance for her husband casts the two of them as loyalists in different modes, the one giving his life to honor his dynasty, the other preserving hers to raise their children.

> Your name will be known forever:
> I have chosen to cling to life.
> Officials who maintain their loyalty are called great,
> Parents who cherish their children, merely human.
> You were a righteous official in life;
> Your epitaph carries your name beyond death.
> Though the living and the dead walk on different roads,
> With my chastity and your integrity, we walk hand in hand.[47]

Such acts of sacrifice were remembered against a background of foreign conquest that could not really be resisted by anything but remorse. China had been through this before. Like the people of the Song, the people of the Ming saw themselves caught in an unbridgeable ethnic gap with their conquerors. The Manchus too were invaders from the steppe, yet they did not choose to rule Mongol-style. The Yuan had accentuated ethnic distinctions to impose order; the Qing preferred the fiction of multinational unity. The reality was a foreign aristocracy whose qualification for rule was brute conquest. The Ming idea of a Chinese China survived as an ideal that excluded steppe customs. It was too firmly settled in Ming minds for them to regard the Manchus as anything other than interlopers from beyond the pale of civilization.

That too would change. Once it was clear that the Manchus were not about to reorganize the realm in any significant way, the social order that had prevailed under the Ming simply resumed. The people of the Ming became, almost seamlessly once the fire of resistance burned out, the people of the Qing. When a republic emerged from the ruins of the empire in 1912, the Ming was fondly recalled as the last "Chinese" dynasty, but what "China" had become no longer fit within the borders that the Ming state had created. The revolutionaries who founded the Republic were uninterested in going back to Ming borders. They claimed sovereignty over all the territory that the Manchus had unified, from Taiwan to Tibet. But then they were only doing what Khubilai Khan and Zhu Yuanzhang had claimed in their turn to have done: unifying the realm. The Yuan and the Ming were not forgotten after all.

CONCLUSION

CHINGGIS Khan's ambition was to conquer the world. His grandson Khubilai set himself the more modest goal of ruling all of eastern Asia. Though neither succeeded entirely, each commanded a world-empire stretching far beyond the Mongols' original homeland. Chinggis's world-empire restlessly pressed onward, eliminating or absorbing the smaller polities it conquered like beads on a string. Indifferent to his cousins' challenges in the less productive western end of his grandfather's empire, Khubilai let the west drop away from his direct control and instead devoted his resources to the conquest of the infinitely more profitable east: the Song dynasty and beyond it Korea, Vietnam, and unsuccessfully Japan. Rather than absorb China into Mongolia, Khubilai led the Mongols into China and assumed his place in the long line of families who had ruled that empire since 221 BC. His polity would be more than tribal beads on a string. It would be a dynastic state.

The logic of empire is political: the expansion of sovereignty for the glory of the ruler. It has no inherent economic logic beyond keeping supporters fed. A world-empire will use military force to collect tribute from the "world" it has brought into being, but it does not exist to secure revenue. The Yuan was different—because of the agrarian realm into which it moved. What kept the Yuan on its feet was a hybrid fiscal regime that drew on both the nomadic tradition of tribute and the administrative tradition of agricultural taxation. Indeed, had the ruling family been able to sort out a more stable system for imperial succession, it might have lasted longer than the century it did.

The Ming dynasty emerged once environmental conditions turned benign in 1368. The new regime rejected everything about the Yuan dynasty except its political constitution and its claim of unification. It continued to maintain the appearance of a world-empire, propped up by the necessary fictions of the tribute system. By returning to pre-Mongol borders and disdaining the steppe as a zone fundamentally alien to Chinese traditions and interests, the Ming relinquished the pose of a world-empire. Nor, though, did it become a world-economy. Its regional economies certainly interacted, and did so increasingly as internal trade expanded through the sixteenth century, but the natural barriers of topography and distance would have kept these regions apart were it not for the state. The strength of administrative practices imposed from a political center gave the Ming the framework for internal integration. This is why the Ming is better conceived as a state-economy than a world-economy.

What drove the Yuan toward empire and the Ming away from it had something to do with their distinctive cultural and political traditions rooted in nomadism and agriculture, but it had much to do as well with changes in the wider world. In the late thirteenth and fourteenth centuries, a continental world-economy oriented the Yuan westward across the grasslands to Persia and Europe.[1] In the sixteenth and seventeenth centuries, a maritime world-economy centered on the South China Sea tied the Ming to systems of trade that flowed to and from the Indian Ocean and across the Pacific. These were different worlds engaging China in different ways.

This shift took place in the context of climate changes that the Yuan and Ming shared with the rest of the world. The weather on its own does not explain the rise of the Yuan or the fall of the Ming, still less everything that occurred between the founding of the one and the demise of the other. But the history of these four centuries cannot be fully understood without taking into account the pressure of weather on society and the state, and more particularly on the economic foundation on which the realm rested, agriculture. Yuan and Ming farmers did not remain the passive victims of climate anomalies, however. By the thirteenth century they had amassed a body of extraordinarily detailed knowledge about how to produce food under conditions as diverse as the arid northern grasslands and the semitropical south. Through practice and adaptation, agricultural knowledge in China had achieved a high tolerance for geographical variation—between north and south, certainly, but also between one province and the next, even one county and the next. Everyone understood that what you could grow in one place was not what you could grow in another.

The capacity of Chinese agriculture to tolerate variation, even to flourish under it, is shown through the enormous number of rice strains grown across the country. Each was developed in relation to local conditions, and all changed over time. As the anthropologist Francesca Bray notes in her history of Chinese agriculture, farmers selected among rice varieties according to those characteristics that would ensure maximum yields. Without this work, rice would not have spread as it did during the Yuan and Ming to become nearly universal throughout the realm, even in the north, traditionally a millet ecology.[2] A scholar of the early Qing collected over three thousand names of rice varieties, and Bray suspects that this figure falls short of the actual number of strains in use. Agricultural knowledge endlessly adapted to changes in local ecologies.

Tolerance for variation in space did not, however, translate easily into tolerance for sudden changes in time. The limits of adaptation were exposed during the worst sloughs, when conditions swung beyond the normal range of oscillation. Violent alterations in climate from year to year undermined the security that the precise adaptation of rice strains was intended to provide. We know too little of climatic conditions before the Yuan to say whether these variations swung harder and wider than they had before, but it appears that they did. A sign of this stress is the appearance of agricultural handbooks and famine administration manuals starting in the fourteenth century. Wang Zhen's *Agricultural Manual,* published in movable type in 1313 and much republished and imitated thereafter, aimed to provide an account of agricultural technologies north and south so comprehensive—the word *bei,* "complete," caps his brief preface—that an official would have in one book all the knowledge he needed to nourish the people.

The handbook includes a monthly calendar of agricultural activities, arranged as a pie chart of twelve segments. The earnest official had only to turn this mandala to the correct month to know what farmers should do, and what the magistrate had to ensure got done. On the one hand, here was a gauge for the efficient application of agricultural knowledge almost mechanical in its completeness.[3] On the other hand, should the weather shift off its usual annual pivot, the mandala could become useless: centuries of fine adjustments negated by a new climate regime, and massive starvation following as a result. The stream of agricultural handbooks, famine manuals, and famine pharmacopoeia that flowed from the pens of well-meaning princes and officials, especially during the last century of the Ming, suggests that the attempts to revise existing knowledge were never entirely successful. Better knowledge was needed, yet what could be added to what farmers had already spent centuries perfecting? To change any component of such knowledge, especially when an ever larger population was working the land ever more intensively, was to run too great a risk.[4] Adaptability had ended in fragility.

The people who lived through the Wanli and Chongzhen sloughs may have been trapped in a deficit of agricultural knowledge, but they were also experiencing an extraordinary recalibration of the local and the global. The growing world-economy of the South China Sea was moving the Ming economy offshore, reorganizing its prices in relation to supply and demand in South America, South Asia, and Europe and no longer just in the domestic market, however large it was. New ideas were also adding to the perplexity. Every new puzzle compounded the old ones to a degree that even the best statecraft minds of the age were baffled to reorient the whole system. Had it not been for the sudden realignment of world-empires with the rise of the Qing dynasty in 1644, this bafflement could have spelled the end of more than just the Ming. Instead, the Manchus shut the borders, replaced the emperor with a khan, and revived the ambitions of empire.

Out of this crucible of political shifts, southern oscillations, and maritime expansion emerged what historians have called the early modern world: a period when growing trading networks inspired innovation and linked separate world-economies into what would become a single global economy. We are used to thinking of people from certain coastal areas of Europe creating this early modern world, but people of the Ming were as much a part of this process as any of the other agents that nursed the system into being.

And then paths diverged. The decade in which the Ming fell to the Qing was also the decade in which European diplomats met at a series of conferences to end the longest-running wars in Europe and consolidate the new forms of political and commercial power that set the lines along which the modern world would develop. The resulting accords, known as the Peace of Westphalia, established the norms of state sovereignty underpinning the world order today. They made states the chief actors in the world system, recognized that every state enjoys a sovereignty that is inviolable, and forbade states from intervening in one another's affairs. The state was no longer the private domain of the monarch but a public entity, no longer a consumer of tribute but an agency to concentrate resources and deploy them entrepreneurially for national ends.[5] The Peace of Westphalia gave the better resourced states of Europe the security they needed to launch new empires that no one would confuse with the "old" empires of the Mongols or the Manchus. Westphalia confirmed that Chinese and European states thereafter were on different courses. State ventures that the Dutch lawyer Hugo Grotius (1583–1645) earlier defended as "the freedom of the seas" Chinese judges prosecuted as "maritime banditry." Even so, Chinese manufacturers and traders continued to provide commodities of a value high enough, and a price low enough, to hold up China's corner of global trade right through the eighteenth century.

To credit Europeans with creating the early modern world singlehandedly ignores the fabric of pre-existing commercial networks into which they wove themselves, and the producers who provisioned the trade, and indeed their own awareness that change was afoot. Zhang Xie was dimly aware of what was happening. Standing on the wharves of Moon Harbor, he could look out into the maritime world and see that a new world was coming into being, one that conformed to different rules and even demanded different personalities. "Once out of Moon Harbor," as we read in his *Study of the Eastern and Western Seas,* "there are no coastlines to follow, no villages to note, and no courier stages to tick off."[6] Unboundedness was not otherwise a condition of Ming life. Those with no experience of seafaring could not but regard the ocean as a chaotic space of danger and disorder, but Chinese mariners—who numbered in the hundreds of thousands in Zhang's day—were learning otherwise. Seven out of ten families in Moon Harbor, Zhang reports, "were familiar with foreigners" and did not fear to sail the eastern and western sea routes to do business with them.

Commercial accumulation, cutthroat competition, conspicuous consumption, a restless rejection of norms and traditions: these were altering social practices and attitudes in China as well as Europe, weaving both places into a common historical process we now call globalization. Call that time the Renaissance, the late Ming, or the early modern world, and all you are doing is switching codes. Each code makes partial sense of the past and present—which is why we will continue to switch them in the future. There will always be more to understand, and more ways to understand. New ways of seeing will not alter the object of our understanding, but they do change its scope, as a Chinese poet once observed:

> You can see the white sun setting behind the mountains,
> and the yellow river disappearing into the ocean.
> But if you wish to see more, you must climb higher:
> then you can see the white sun setting behind the mountains,
> and the yellow river disappearing into the ocean.[7]

This history started with dragons, so let it end with two, one we can see and one we can't. The first appears to us in a scroll painting of a *lohan* or Buddhist saint seated in meditation (Fig. 18). Wu Bin, a professional painter who worked at the court of the Wanli emperor and was active throughout his reign, painted it in 1601. Compared to the dragons with

Fig. 18 *Lohan* by Wu Bin, 1601. The subject—a dragon appearing at a charged moment—is traditional, but the use of shading to convey cylindrical surfaces betrays the influence of European art. The first European engravings began to circulate in China at the turn of the seventeenth century, and Wu must have seen them. National Palace Museum, Taiwan, Republic of China.

which Ming viewers were familiar, this one looks a little bizarre. The tiny head accentuates the snakelike character of its scaly body. Notice also how the light strikes it. Wu Bin has lit both sides of the body and shaded the convex surface between the sides. It is a device that Italian artists had recently developed, known as chiaroscuro: the use of shadow and light to portray volume in three-dimensional objects. This was not a technique Chinese artists used. Chiaroscuro is also evident on the strange geometrical rocks crowding around the Buddhist monk, as well as on the neat columns of tree trunks rising behind his left shoulder and the leafy branches botanically edged in black ink. And what are we to make of the thick white clouds on which the dragon descends? Far more like Italian stonework than the airy mists at which Chinese artists had excelled for centuries.

What we have here, then, is a Chinese painting that Ming viewers would not have recognized as a "Chinese" painting. It looks utterly Chinese to us, but Wu Bin and his dragon are crossing cultures. Wu Bin was not consciously trying to imitate a European style, yet that style has leaked through, entering Wu's visual imagination to fuel his own creative originality. The art historian James Cahill made the discovery, attributing it to the arrival of European engravings brought by Jesuit missionaries and disseminated in local woodblock reproductions.[8] Wu has seen European dragons—perhaps the serpent in the Garden of Eden?—and has added them to the Chinese repertoire.

Now for our final dragon, the one we can't see. It is the last dragon whose appearance can be precisely dated to the Ming dynasty: September 26, 1643.[9] It was a great shining creature that rose up in the night sky over the southeastern hills of Shanxi province. Nothing heralded its coming, neither a shred of cloud nor a whisper of thunder. Suddenly it was there in the sky, twisting aloft in the pure white moonlight. Its body emitted a glow that poured golden light through the doors and windows of the houses below, awakening the sleepers. Everyone stepped outside and gazed up in awe at this glorious and peaceful sight. No one could guess what it meant or what it portended. How could they see what was coming, when neither can we?

TEMPERATURE AND PRECIPITATION EXTREMES

THE NINE SLOUGHS

SUCCESSION OF EMPERORS

PRONUNCIATION GUIDE

NOTES

BIBLIOGRAPHY

ACKNOWLEDGMENTS

INDEX

PERIODS OF TEMPERATURE AND PRECIPITATION EXTREMES, 1260–1644

Temperature		*Precipitation*	
Cold	1261–1393	Dry	1262–1306
		Wet	1308–25
		Dry	1352–74
		Wet	1403–25
Cold	1439–1455	Dry	1426–1503
Warm	1470–76		
Cold	1481–83		
Cold	1504–09		
		Dry	1544–1643
Warm	1536–71	Severe	1544–46
Cold	1577–98	Severe	1585–89
Cold	1616–20	Severe	1614–19
Severe	1629–43	Severe	1637–43

THE NINE SLOUGHS OF THE YUAN AND MING DYNASTIES

Years	*Reign era*	*Conditions*
1295–97	Yuanzhen	Drought, flood, dragon
1324–30	Taiding	Drought, famine, locusts
1342–45	Zhizheng	Cold, drought, famine, flood, epidemics
1450–55	Jingtai	Cold, wet, famine, flood, epidemics
1516–19	Zhengde	Cold, wet, famine, earthquake, epidemics, dragons
1544–46	Jiajing	Cold, drought, famine, epidemics
1586–88	Wanli I	Cold, drought, famine, flood, locusts, epidemics, dragons
1615–17	Wanli II	Cold, drought, famine, locusts, earthquake, dragons
1637–43	Chongzhen	Cold, drought, famine, locusts, earthquake, epidemics, sandstorms, dragons

THE SUCCESSION OF THE YUAN AND MING EMPERORS

Personal name	*Reign titles*	*Inaugural year*	*Relationship to predecessor*
Yuan Dynasty 1271–1368			
1. Khubilai	Zhiyuan	1271	Nephew of Ögödei, grandson of Chinggis
2. Temür	Yuanzhen Dade	1294	Youngest grandson
3. Khaishan	Zhida	1308	Nephew (son of Temür's elder brother Darmabala)
4. Ayurbarwada	Huangqing Yanyou	1311	Younger brother
5. Shidebala	Zhizhi	1321	Son
6. Yesün Temür	Taiding	1323	Uncle (son of Temür's eldest brother Kammala)
	Zhihe		
7. Aragibag	Tianshun	1328	Son
8. Tugh Temür	Tianli	1328	Cousin (son of Khaishan, grandson of Darmabala)
9. Khoshila		1329	Elder brother
0. Tugh Temür	Zhishun	1329	Younger brother

11. Irinjibal		1332	Nephew (son of Khoshila, great-grandson of Darmabala)
12. Toghön Temür	Yuantong Zhiyuan Zhizheng	1333	Elder brother

Ming Dynasty 1368–1644

1. Yuanzhang	Hongwu	1368	
2. Yunwen	Jianwen	1398	Grandson (son of Yuanzhang's eldest son)
3. Di	Yongle	1402	Uncle (fourth son of Yuanzhang)
4. Gaozhi	Hongxi	1424	Eldest son
5. Zhanji	Xuande	1425	Eldest son
6. Qizhen	Zhengtong	1435	Eldest son
7. Qiyu	Jingtai	1449	Younger half-brother
8. Qizhen	Tianshun	1457	Elder half-brother
9. Jianshen	Chenghua	1464	Eldest son
10. Youtang	Hongzhi	1487	Eldest surviving son
11. Houzhao	Zhengde	1505	Only son
12. Houcong	Jiajing	1521	Younger cousin (grandson of Jianshen)
13. Zaihou	Longqing	1567	Eldest surviving son
14. Yijun	Wanli	1572	Eldest surviving son
15. Changle	Taichang	1620	Eldest son
16. Youjiao	Tianqi	1620	Eldest son
17. Youjian	Chongzhen	1627	Younger brother (fifth son of Changle)
18. Changxun	Hongguang	1644	Cousin (grandson of Yijun)

PRONUNCIATION GUIDE

c	as *ts* in *nets*
ch	as in *chat*
g	as in *girl*
j	as in *jingle*
q	as *ch* in *cheese*
x	as *sh* in *sheer*
y	as in *year*
z	as *dz* in *adze*
zh	as *j* in *John*
a	as *e* in *pen* for yan, jian, qian, xian; otherwise as *a* in *father*
ai	as in *aye*
ang	as *ong* in *wrong*
ao	as *ow* in *now*
e	as *e* in *yet* in the combinations ye, -ie, -ue; otherwise as *e* in *the*
ei	as in *neigh*
en	as *un* in *fun*
eng	as *ung* in *rung*
er	pronounced as *are*
i	as in the *i* of *sir* after c, s, z; as in the *ir* of *sir* after ch, sh, zh, r
ie	as *ye* in *yet*
iu	as *yo* in *yoyo*
ong	as *ung* in German *Achtung*
ou	as in *oh*
u	after j, q, x, and y as *ui* in *suit;* otherwise as *u* in *rule*
ua	after j, q, x, and y as *ue* in *duet;* otherwise as *wa* in *water*
uai	as in *why*
ue	as *ue* in *duet*
ui	as in *way*
uo	similar to *o* in *once*

NOTES

1. Dragon Spotting

1. This chapter is based on close to 100 dragon sightings culled from the dynastic histories of the Yuan and Ming (Song Lian, *Yuan shi,* 1099; Zhang Tingyu, *Ming shi,* 439–440), local gazetteers, and commonplace books. The 1293 sighting, recorded in *Haiyan xian tujing* (1624), 3.54a–55b, and reprinted in *Jiaxing fuzhi* (1879), 11.6a, is translated in Elvin, *The Retreat of the Elephants,* 196.

2. Zhang Tingyu, *Ming shi,* 439. The gazetteer of Linqu county records a meteor striking the hill in July 1363, but no other unusual event during that decade; *Linqu xianzhi* (1552), 1.8b, 4.20b, 4.28b. The provincial gazetteer, *Shandong tongzhi* (1533), 39.36b, confirms the Linqu account.

3. Tao Zongyi, *Nancun chuogeng lu,* 105. I am grateful to Desmond Cheung for alerting me to this passage.

4. Zhu Yuanzhang, *Ming taizu ji,* 350–351.

5. For example, see the comments of Jiao Hong, *Yutang congyu,* 109–110.

6. Zhang Yi, *Yuguang jianqi ji,* 1025.

7. Sterckx, *The Animal and the Daemon in Early China.*

8. This general rule is contradicted by the *History of the Qing,* which records the dynasty's first dragon sighting in 1649, barely five years into the new regime; Zhao Erxun, *Qing shi gao,* 1516. Was this sighting recorded to cast doubt on the Manchus' right to rule?

9. *Ming wuzong shilu,* 150.3a, 162.2b.

10. Huang, *1587, a Year of No Significance,* provides a sharply critical account of the Zhengde emperor.

11. "The Emperor's return north after getting ill as a result of falling into the water is in reality a direct response to the earlier events of [dragons] sucking up

boats and releasing floods"; Shen Defu, *Wanli yehuo bian,* 742. To the Zhengde-era dragon sightings in the *Ming History,* Shen adds another two, one of which he appears to have taken from Lu Can (1494–1551), who includes a slightly longer version of the same story in his *Gengsi bian,* 105. Shen precedes his essay on Zhengde dragon anomalies with a longer essay on other "odd transformations" during the Hongzhi reign.

12. Zhang Yi, *Yuguang jian qiji,* 1024.

13. Topsell, *The Historie of Serpents,* 155, 161–162. The spelling of some words has been modified. My thanks to Keith Benson for introducing me to Topsell.

14. Chen Yaowen, *Tianzhong ji,* 56.10a, 19b, 20a.

15. Lu Rong, *Shuyuan zaji,* 14. He makes a similar comment in another passage, in which he discusses a subspecies of smaller dragon known as *jiao,* which he says lacks the capacity of a full dragon to self-transform (185). Paul Smith draws extensively on Lu Rong's writings in his "Impressions of the Song-Yuan-Ming Transition," especially 95–110.

16. Topsell, *The Historie of Serpents,* 153.

17. Chen Yaowen, *Tianzhong ji,* 56.2b.

18. Lang Ying, *Qixiu leigao,* 289.

19. Lang Ying, *Qixiu leigao,* 645; I am grateful to Desmond Cheung for alerting me to Lang's essays on dragons.

20. Lu Rong, *Shuyuan zaji,* 154.

21. Xie Zhaozhe, *Wu zazu,* 166–167.

22. Topsell, *Historie of Serpents,* 172–173.

23. Ye Ziqi, *Caomuzi,* 16. For Ming understandings of the medical properties of dragons, see Nappi, *The Monkey and the Inkpot,* 55–68.

24. Tan Qian, *Zaolin zazu,* 483; *Shanxi tongzhi* (1682), 30.40b. The term "dragon bones" was also used for the oracle bones that royal priests used for divination in the Shang dynasty and then buried, which materia medica collectors also exhumed. Both types of bone continued to be harvested and ground up for medicine into the twentieth century; see Andersson, *Children of the Yellow Earth,* 74–76; Schmalzer, *The People's Peking Man,* 35–37, 132–134.

25. Mark Elvin is the sole modern historian to suggest that sightings of "superfauna" such as dragons be treated seriously as historical evidence of how people at the time saw the world; see his *The Retreat of the Elephants,* 370.

26. Gould, "Foreword," xiv.

27. Zhao Erxun, *Qing shi gao,* 1519.

28. Hurn, "Here Be Dragons? No, Big Cats," 11. For bringing this article to my attention, I am grateful to Gustaaf Houtman, who published it in *Anthropology Today.*

29. Li Qing, *Sanyuan biji,* 153.

2. Scale

1. Polo, *The Travels,* 113.

2. There were others traveling in the opposite direction who might have enlarged the European vision, such as the Chinese Nestorian monk Rabban Sauma, who left Beijing in 1275 and met the kings of France and England in 1287, but none of their writings were translated into European languages. Sauma's story is told in Rossabi, *Voyager from Xanadu.*

3. Polo, *The Travels,* 113, 125, 129, 130.

4. Coleridge in "Kubla Khan" imagines the Mongol ruler's summer residence in Shangdu (Xanadu). His images draw heavily on reports from Mughal India. Mongolia had no "incense-bearing trees" or "forests ancient as the hills," no "deep romantic chasms" shrouded in cedars. As for a sacred river running "through caverns measureless to man down to a sunless sun," the Luan River does flow to the ocean four hundred kilometers away, but there is not a cavern in sight. How curious that this poem, which Coleridge wrote while taking opium as a palliative for illness, is where English-language schoolchildren make their first acquaintance with the founder of the Yuan.

5. Wood, *Did Marco Polo Go to China?* 96. Though I disagree with Wood's conclusion, I recommend the book as a delightful introduction to the confusions of Polo's record and the complications of his world.

6. Polo, *The Travels,* 85, 91.

7. Waldron, *The Great Wall of China,* 140–164.

8. Needham, *Science and Civilisation in China,* V:6, 219–225.

9. Delgado, *Khubilai Khan's Lost Fleet.*

10. Zicong (1216–1274) is now better known by the lay name that Khubilai later gave him, Liu Bingzhong. See Hok-lan Chan's biography in de Rachewiltz et al., *In the Service of the Khan,* 245–269.

11. E.g., Liu Ji, *Da Ming qinglei tianwen fenye zhi shu,* preface, 6a; Zhu Yuanzhang, *Ming taizu ji,* 9; Huang Yu, *Shuanghuai suichao,* 12; *Ming taizu shilu,* 56.11b.

12. *Ming taizu shilu,* 56.12a.

13. Zhang Yi, *Yuguang jianqi ji,* 120.

14. Wang Qi, *Sancai tuhui,* 1.7a–b.

15. Scott, *The Art of Not Being Governed,* 12.

16. The Nine Frontiers refers to the nine regional military commissions stretching from Liaodong Command in the far northeast to Gansu Command in the far northwest. The other seven from east to west were Jizhou, Xuanfu, Dadong, Taiyuan, Yansui, Shaanxi, and Ningxia.

17. Wang Shixing, *Guangzhi yi,* 2.

18. Polo, *The Travels,* 150–154. The translator muddles the two systems when on 151 he faults Polo for "some confusion about the foregoing figures." Polo was

in fact correct in distinguishing the two systems: the stations twenty-five miles apart are courier stations *(yi),* and those three miles apart are postal stations *(pu).* Polo is remarkably accurate with his figures. Courier stations were supposed to be 60 *li* apart (35 km, 22 miles), and postal stations 10 *li* (6 km, 3½ miles); see Brook, "Commerce and Communication," 582, 594.

19. *Cili xianzhi* (1574), 6.12b.

20. Li Le, *Jianwen zaji,* 1.18b.

21. Shen Dingping, quoted in Brook, *The Confusions of Pleasure,* 35.

22. *Jing'an xianzhi* (1565), 1.18a.

23. Jiang, *The Great Ming Code,* 146.

24. *Da Yuan shengzheng guochao dianzhang,* 36.6b–8a.

25. The time limits are from Ye Shiyong's enlarged 1586 edition of Tao Chengqing, *Da Ming yitong wenwu zhusi yamen guanzhi,* a printed guide to the Ming administrative system. A unique copy survives in Beijing because a grand coordinator in Jiangxi happened to forward a copy to the court. Provincial averages may be found in Tong, *Disorder under Heaven,* 129. Tao's name is associated with the popular route book, *Shangcheng yilan* (Merchant routes at a glance); see Brook, *Geographical Sources of Ming-Qing History,* entry 4.1.2.

26. The distance from Guangzhou to Chaozhou has been calculated from the route data in Yang Zhengtai, *Tianxia shuilu lucheng,* 88.

27. Xie Zhaozhe, *Wu zazu,* 4.16b. For a brief introduction to this commonplace book, see Oertling, *Painting and Calligraphy in the Wu-tsa-tsu,* 1–4.

28. Wang Shixing, *Guangzhi yi,* 3.

29. Chen Quanzhi, *Pengchuang rilu,* 1.38a–b. Chen also notes another environmental sign of difference: "the north has a lot of tree borers but no centipedes, and the south has a lot of centipedes but no tree borers," while in only certain prefectures in the Huai valley "do both creatures reproduce." On the division between rice and millet agriculture, see Brook, *The Chinese State in Ming Society,* 81–83.

30. Wang Daokun, *Taihan ji,* 494.

31. Wang Shixing, *Guangzhi yi,* 2–3.

32. Wang Shixing, *Guangzhi yi,* 5.

33. This discussion is based on Elman, *A Cultural History of Civil Examinations,* 90–97.

34. Zhang Tingyu, *Ming shi,* 7344; Zhang Yi, *Yuguang jianqi ji,* 1025. On tributary students, see Dardess, *A Ming Society,* 160–166.

35. These were Lingbei (Outer Mongolia and part of Siberia), Liaoyang (Manchuria and northern Korea), and Zhendong (southern Korea, a region that remained effectively under Korean control and kept the Mongols at bay through a tributary arrangement).

36. On Ming administrative geography, see Guo and Jin, *Zhongguo xingzheng quhua tongshi: Mingdai juan;* also Hucker, *A Dictionary of Official Titles,* 62–65,

75–78. With this reorganization, today's administrative map of China was largely set. The only significant change in the Qing was to divide Huguang into North (Hubei) and South (Hunan). So too the names of cities today largely follow the names with which the Ming replaced Yuan usages: Jiqing became Nanjing, Bianliang Kaifeng, Fengyuan Xi'an, Jingjiang Guilin, Shunyuan Guiyang, and Zhongqing Kunming.

37. Sedo, "Environmental Jurisdiction," 8; see also Nimick, *Local Administration,* 79–82. Des Forges, *Cultural Centrality,* 22–66.

38. Jiangxi province, underpoliced and overpopulated, got 7 new counties in the mid-Ming; Zhang Tingyu, *Ming shi,* 1057–1067.

39. Shi Ru, "Qing fenli xianzhi shu," quoted in Liu Shiji, "Ming Qing shidai Jiangnan diqu de zhuanye shizhen," 1.

40. *Haicheng xianzhi* (1762), 21.1a–4a.

41. Song Lian, *Yuan shi,* 1345. The population figures given in this chapter are from Liang, *Zhongguo lidai hukou,* 176ff.

42. Li Xu, *Jiean laoren manbi* (Random notes by the Old Man of Austerity Hermitage), quoted in Cao Shuji, *Zhongguo renkou shi,* 19; also in Li Defu, *Mingdai renkou yu jingji fazhan,* 24; translated differently in Ho, *Studies,* 4–5.

43. Zhang Xuan, *Yi yao* (Doubts and clarities), quoted in Li Defu, *Mingdai renkou yu jingji fazhan,* 26.

44. *Kaizhou zhi* (1534), 3.3a.

45. *Lanyang xianzhi* (1545), 2.8b.

46. The moderate skeptics are represented by Ho Ping-ti (*Studies on the Population of China,* 22) and Cao Shuji (*Zhongguo renkou shi*), the ultra-skeptics by Martin Heijdra ("The Socio-economic Development of Rural China during the Ming"), and the fundamentalists by Li Defu (*Mingdai renkou yu jingji fazhan,* 48–54). For an overview of some of the controversies surrounding this issue, see Marks, "China's Population Size during the Ming and Qing."

47. Zhang Qing, *Hongdong dahuaishu yimin zhi,* 55.

48. *Heze Wang shi jiapu* (Genealogy of the Wang family of Lotus Marsh) (1887), Wang Mingluan's preface, excerpted in Zhang Qing, *Hongdong dahuaishu yimin zhi,* 97–98. The date of the 1370 relocation order may be in error; as Zhang Qing notes elsewhere in his book (48), the earliest recorded relocation was not until 1373.

49. Liang, *Zhongguo lidai hukou,* 205–207. For estimates of provincial densities, see Li Defu, *Mingdai renkou yu jingji fazhan,* 111–112.

50. Yu Jideng, *Diangu jiwen,* 183.

51. *Huguang tujing zhishu* (1522), 1.66b.

52. Brook, *The Chinese State in Ming Society,* 22–32.

53. Although Watertight Registers were not mandated in the Ming, they continued to be compiled in some counties to provide cross-checks on the Fish-Scale registers; see Hai Rui, *Hai Rui ji,* 160, 190–192, 285–287.

54. Lu Rong, *Shuyuan zaji,* 84.

55. Hai Rui, *Hai Rui ji,* 159, 190–198. For a detailed account of county mapping for the Great Compilation year of 1572, see Brook, *The Chinese State in Ming Society,* 43–59.

56. Fuzheng, *Hanshan dashi nianpu shuzhu,* 46.

57. This variation is explored in Nimick, *Local Administration.*

3. The Nine Sloughs

1. *Qiongshan fuzhi* (1618), 12.3a. The typhoon is dated August 15; the date of the dragon attack is not given.

2. *Qiongshan fuzhi* (1618), 12.1b–12b.

3. Shen Jiaben, *Shen Jiyi xiansheng yishu,* vol. 2, 7.8a-b; *Qiongzhou fuzhi* (1890), 31.3b-4a. The prefect, Wu Hui, was appointed in 1657 and served as an exemplary administrator for five years.

4. Song Lian, *Yuan shi,* 1051–1115; Zhang Tingyu, *Ming shi,* 427–512. The latter introduces a few categories not found in the former, such as a chronology of rat infestations between 1616 and 1644 (477). The reconstruction I have done on the basis of these histories is experimental, merely a first approximation using these data of an aspect of history we have so far neglected.

5. Lu Rong, *Shuyuan zaji,* 81–82; on Zhou's local reputation, see p. 59.

6. *Songjiang fuzhi* (1630), 47.19b–20a, 21b.

7. *Qiongshan fuzhi* (1618), 12.1a.

8. This final period is known as the Maunder Minimum, named after the English astronomer Edward Maunder (1851–1928), who correlated the lower temperatures with the virtual disappearance of sunspot activity.

9. Grove, "The Onset of the Little Ice Age," 160–162.

10. Zhang and Crowley, "Historical Climate Records in China," 841.

11. Zhang Yuniang, "Singing of Snow," translated by Anna Marshall Shields, in Chang and Saussy, *Women Writers of Traditional China,* 149.

12. Sedo, "Environmental Jurisdiction," 5.

13. *Jiangdu xianzhi* (1881), 2.13b.

14. Gallagher, *China in the Sixteenth Century,* 316.

15. For other snow paintings by Dai Jin, see Cahill, *Parting at the Shore,* 15; Wu, *Orchid Pavilion Gathering,* figs. 6, 8, 9; Li and Knight, *Power and Glory,* pl. 119; Gao, *Paintings of the Ming Dynasty,* pl. 4.

16. For snow paintings by Tang Yin and Zhou Chen, see Clapp, *The Painting of T'ang Yin,* figs. 52, 60 and 65.

17. For other snow paintings by Wen Zhengming, see Clunas, *Elegant Debts,* pl. 22, 74; idem., *Empire of Great Brightness,* pl. 29.

18. Cahill, *Parting at the Shore,* 29.

19. For snow paintings by Zhao Zuo, see Cahill, *Compelling Image,* 82; Gao, *Paintings of the Ming Dynasty,* pl. 65.

20. The cold and snow continue until March 25, when "the weather turned

muggy, things are mildewed and running with condensation, and the birds call without stop." There is one more cold snap on April 12 ("cold in the extreme") before temperatures return to normal; Li Rihua, *Weishui xuan riji,* 495–519.

21. The Palace Museum in Beijing holds three snow paintings by Zhang Hong.

22. Readers interested in a year-by-year summary of precipitation may wish to consult the annual precipitation maps compiled by the Central Hydrological Bureau in Beijing based on reports taken from local gazetteers starting in 1470; Zhongyang qixiangju qixiang kexue yanjiuyuan, *Zhongguo jin wubai nian hanlao fenbu tuji.*

23. Zhang Tingyu, *Ming shi,* 485.

24. Whether the rain comes or fails depends on forces far beyond the locality affected. Meteorological research has uncovered a pattern to deviations from normal rainfall in the tropical and temperate regions of the Pacific Ocean connected to El Niño. El Niño is the warm equatorial current that every four to six years moves up the west coast of South America in winter, bringing high humidity and heavy rain to the Peruvian coast. In Southeast Asia, the effect reverses and the monsoon rains weaken, causing drought and cooler temperatures. El Niño cannot explain the prolonged stretches of dry weather during the Yuan and Ming, which seem to indicate longer-term shifts rooted in the climate of the continent rather than the ocean. Still, some of the severe episodes of drought in the last century of the dynasty correlate with El Niños, most compellingly in the mid-1540s, the late 1580s, and the late 1610s. See Quinn, "A Study of Southern Oscillation-Related Climatic Activity," 126.

25. Zhang Yi, *Yuguang jianqi ji,* 1024.

26. Song Lian, *Yuan shi,* 1053.

27. Song Lian, *Yuan shi,* 1058.

28. Guojia dizhenju diqiu wuli yanjiusuo, *Zhongguo lishi dizhen tuji: yuangu zhi Yuan shiqi,* 151–156; Gu Gongxu, *Catalogue of Chinese Earthquakes,* 19–21.

29. Lu Can, *Gengsi bian,* 105.

30. Gu Gongxu, *Catalogue of Chinese Earthquakes,* 44–52.

31. Xie Zhaozhe, *Wu zazu,* 4.17a.

32. Gu Gongxu, *Catalogue of Chinese Earthquakes,* 67–69.

33. The dates are taken from Lentz, *The Volcano Registry.*

34. Benedictow, *The Black Death,* 50.

35. Benedictow, *The Black Death,* 18, 26, 49–51, 229–231, 235. The data given are for bubonic plague (transmitted by flea bite), not pneumonic plague (transmitted directly via water droplets from the breath of the infected person to the lungs of another); Benedictow argues that pneumonic plague was rare during the Black Death. Li Bozhong, "Was There a 'Fourteenth-Century Turning Point'?" 138, is similarly skeptical of the claim that the 1344 epidemic was plague.

36. Cao Shuji's hypothesis is summarized and supported in Hanson, "Invent-

ing a Tradition in Chinese Medicine," 97–102. The 1582 and 1587 epidemics in Beijing are noted in Zhang Tingyu, *Ming shi,* 443.

37. Xie Zhaozhe, *Wu zazu,* 26, retranslated from Dunstan, "The Late Ming Epidemics," 7.

38. Quoted in Benedictow, *The Black Death,* 4.

39. *Ming shenzong shilu,* 186.2a.

40. Hanson, "Inventing a Tradition in Chinese Medicine," 109.

41. Tan Qian, *Zaolin zazu,* 280.

42. *Haiyan xianzhi* (1876), 13.5a.

43. *Ming xiaozong shilu,* 65.5a.

44. *Shaoxing fuzhi* (1586), 13.32b.

45. The great famine of 1588 remains virtually unstudied; a preliminary survey is included in Dunstan, "The Late Ming Epidemics," 8–18.

46. *Shenzong shilu,* 188.4a.

47. *Shenzong shilu,* 197.3a, 197.11a, 198.2a.

48. Zhang Tingyu, *Ming shi,* 1.

49. Deteriorating conditions between 1434 and 1448 are summarized by Twitchett and Grimm in *The Cambridge History of China,* vol. 7, 310–312, though their account stops short of the Jingtai Slough.

50. Elvin, "Who Was Responsible for the Weather? Moral Meteorology in Late Imperial China."

51. Zhang Tingyu, *Ming shi,* p. 5503.

52. *Ming xiaozong shilu,* 84.2b–4a.

53. *Cili xianzhi* (1574), 6.4a–6a. For a similar set of prognostications for Shanghai, see *Shanghai xianzhi* (1588), 1.10b–11b.

54. Mathematically the Ten Stems and Twelve Branches could generate 120 pairs, but Chinese reckoning restricted the set to half that number.

55. Yu Xiangdou, *Wanyong zhengzong,* 3.4b. Readers interested in other divinations in this encyclopedia can read about them in Brook, *The Confusions of Pleasure,* 163–167.

56. The Lichun dates are from Wang Shuanghuai, *Zhonghua rili tongdian,* 3845–3864.

57. Zhang Tingyu, *Ming shi,* 453, 475.

58. Ye Ziqi, *Caomuzi,* 47.

59. Jiao Hong, *Yutang congyu,* 93.

4. Khan and Emperor

1. On the distinction between khan and great khan, see Allsen, "The Rise of the Mongolian Empire," 332, 367. Marco Polo introduced both terms into European languages; see *The Travels,* 113. For an excellent account of what it meant to be a khan, see Fletcher, "The Mongols: Ecological and Social Perspectives," 21–28.

2. On tanistry, see Fletcher, "The Mongols: Ecological and Social Perspectives," 24–26, 36–38.

3. Ratchnevsky, *Genghis Khan,* 140.

4. de Bary, *Waiting for the Dawn,* 99.

5. Dardess, "Did the Mongols Matter?"

6. Chan, "Li Ping-chung," 252, 258.

7. Rossabi, *Khubilai Khan,* 130.

8. Ye Ziqi, *Caomuzi,* 47.

9. Hsiao, "Mid-Yuan Politics," 531–532.

10. Di Cosmo, "State Formation and Periodization in Inner Asian History," 34.

11. Hucker, *The Ming Dynasty: Its Origins and Evolving Institutions,* 33.

12. Mote, "The Growth of Chinese Despotism," 18.

13. Blue, "China and Western Social Thought in the Modern Period," 86–94.

14. Farmer, *Zhu Yuanzhang and Early Ming Legislation,* 100.

15. Mote, "The Growth of Chinese Despotism," 32.

16. Lu Rong, *Shuyuan zaji,* 123.

17. Brook et al., *Death by a Thousand Cuts,* 116.

18. The *Ancestral Instruction* is translated in Farmer, *Zhu Yuanzhang and Early Ming Legislation,* 114–149; the quotation appears on p. 118.

19. Zhang, *Ming shi,* 7906–7908.

20. Hucker, "Ming Government," 76.

21. For examples of officials who died in the defense of Jianwen, see *Shandong tongzhi* (1533), 25.10b–11a. It was later claimed that Yongle treated well those who lived up to their vow of loyalty and fought for Jianwen; Lu Rong, *Shuyuan zaji,* 28. Other accounts, however, accuse his troops of vicious treatment of the defeated.

22. Zhang, *Ming shi,* 4019; on Fang's conservatism, see pp. 4053–4054.

23. Yu Jideng, *Diangu jiwen,* 107.

24. Church, "The Colossal Ships of Zheng He," 174–175. Church sensibly corrects the historical record on the size of these ships, which were probably a third the length and a tenth the burthen given in the *Ming shi* (160–162). For a sensible description of these famous voyages, see Tsai, *Perpetual Happiness,* 197–208.

25. Geoff Wade in "The Zheng He Voyages: A Reassessment" has argued that Zheng's expeditions be seen as one prong of Yongle's expansion into Southeast Asia, and a continuation of Mongol attempts to move southward. Just as the Mongols imposed their Pax Mongolica over continental Asia, he suggests, so Yongle wished to impose a Pax Ming over maritime Asia. The interpretation draws attention to the large military component on the voyages, but it may misplace the purpose of military power, which was to overawe rather than subdue.

26. Shen Defu, *Wanli yehuo bian,* 9; Yu Jideng, *Diangu jiwen,* 196.

27. Twitchett and Grimm, "The Cheng-t'ung, Ching-t'ai, T'ien-shun Reigns," 323. On the military consequences of this campaign, see Waldron, *The Great Wall*, 87–90.

28. Huang Yu, *Shuanghuai suichao*, 101.

29. Lu Rong, *Shuyuan zaji*, 37; Zhang Tingyu, *Ming shi*, 4411.

30. Twitchett and Grimm, "The Cheng-t'ung, Ching-t'ai, and T'ien-shun Reigns," 339.

31. On Liu Jin, see Geiss, "The Cheng-te Reign," 405–412.

32. Geiss, "The Cheng-te Reign," 433.

33. Waltner, *Getting an Heir*, 1–3.

34. Zhang, *Ming shi*, 5077–5078.

35. Fisher, *The Chosen One*, 72–80, 163–167; Brook, "What Happens When Wang Yangming Crosses the Border?" Wang shows his hand in his third report to the emperor on the Guangxi campaign, in which he praises Jiajing as "the emperor who promotes perfect filial piety to rule the realm"; *Wang Yangming quanji*, 470.

36. Geiss, "The Chia-ching Reign," 450.

37. These events are described in Hsia, *A Jesuit in the Forbidden City*.

38. Wang Xijue, "Quanqing zhenji shu" (Memorial calling for famine relief), in Chen Zilong, *Ming jingshi wenbian*, 395.7b.

39. Wanli's recalcitrance is delightfully portrayed by Ray Huang in the opening chapter of his *1587, a Year of No Significance*.

40. Huang, "The Lung-ch'ing and Wan-li Reigns," 517.

41. Lu Rong, *Shuyuan zaji*, 16. On the compilation of the *Great Encyclopedia*, see Tsai, *Perpetual Happiness*, 133.

42. Wang Daokun, *Taihan ji*, 494–495.

43. The following year, Wanli assigned officials from the Hanlin Academy to go down to the provinces and oversee the provincial examinations in a move to curb provincial independence; Elman, *A Cultural History of Civil Examinations*, 151–152.

5. Economy and Ecology

1. Polo, *The Travels*, 152, 156, 200–201, 204–205, 215, 306.

2. Meskill, *Ch'oe Pu's Diary*, 93–94.

3. Liang Fangzhong, *Zhongguo lidai hukou, tiandi, tianfu tongji*, 303. The fact that the levy twenty-six years later reported collecting exactly the same amount of grain does not inspire confidence. The only sustained study of Yuan economic geography, Wu Hongqi's *Yuandai nongye dili*, avoids any attempt to quantify agricultural production.

4. Ma Wensheng (1426–1510), quoted in Zhang Yi, *Yuguang jianqi ji*, 73.

5. Huang, "The Ming Fiscal Administration," 107.

6. Liang Fangzhong, *Zhongguo lidai hukou, tiandi, tianfu tongji,* 344.

7. Ray Huang proposed that an agrarian empire could not survive by taking less than ten percent; *Taxation and Governmental Finance,* 174, 183. The Hongwu fiscal system appears to have set that minimum bar.

8. Zhu Yuanzhang, "Dagao wuchan xu" (1385), in Yang Yifan, *Da gao yanjiu,* 426. Zhu notes that a family that falls below roughly eighty *mu* will not grow enough to feed itself.

9. Ma Zhibing, "Mingchao tudi fazhi," 421, citing an edict of 1578.

10. Zhang Yi, *Yuguang jianqi ji,* 509.

11. Provincial-level collection and disbursement data are summarized in Liang Fangzhong, *Zhongguo lidai hukou, tiandi, tianfu tongji,* 375.

12. *Ming wuzong shilu,* 158.4b.

13. This principle of Chinese state practice is most clearly laid out in Wong, *China Transformed,* 135–149. Although he bases his argument on evidence from the Qing dynasty, the same principle animated state practice in earlier dynasties.

14. On state textile production in the Ming, see Schäfer and Kuhn, *Weaving and Economic Pattern in Ming Times.*

15. Lu Rong, *Shuyuan zaji,* 66.

16. Wu Jihua, *Mingdai haiyun ji yunhe de yanjiu,* 35–42; Brook, "Communications and Commerce," 596–605.

17. Sedo, "Environmental Jurisdiction," 4.

18. Tan Qian, *Zaolin zazu,* 39–40. On "fast-as-horse boats" *(mahuai chuan),* see Hoshi Ayao, *Min-Shin jidai kōtsūshi no kenkyū,* 88–124.

19. Li Dongyang, "Chongxiu Lüliang hong ji" (Record of restoring the locks at Lüliang Rapids), in Chen Zilong, *Ming jingshi wenbian,* 54.19a, quoted in Feng Yuejian, "Mingdai Jing-Hang yunhe de gongcheng guanli," 50.

20. Deng, *The Premodern Chinese Economy,* doubts that an integrated national economy emerged in the Ming.

21. Reprinted in Yang Zhengtai, *Tianxia shuilu lucheng,* 334–342.

22. Lu Rong, *Shuyuan zaji,* 8.

23. Li Bozhong, "Was There a 'Fourteenth-Century Turning Point'?" 145.

24. On the growth of cotton as China's leading textile through the Yuan and Ming, see Zurndorfer, "The Resistant Fibre," 44–51.

25. Chen Jian, *Huang Ming congxin lu,* 18.18b; Song Lian, *Ming shi,* 946.

26. Yang Zhengtai, "Ming-Qing Linqing de shengshuai yu dili tiaojian de bianhua," 117–119.

27. Jining, 200 kilometers further down the Grand Canal, is another good example of this developmental logic; see Sun, "City, State, and the Grand Canal."

28. Heijdra, "The Socio-Eocnomic Development of Rural China during the Ming," 511.

29. Huang, "Ming Fiscal Administration," 147.

30. *Jingzhou fuzhi* (1880), 4.2b.

31. Tao Zongyi, *Nancun chuogeng lu,* 116.

32. *Nanping xianzhi* (1921), 2.16–17a.

33. Fei, *Negotiating Urban Space,* 1.

34. Zheng Yunduan, "On Husband-Longing Rock," trans. Peter Sturman, in Chang and Saussy, *Women Writers of Traditional China,* 134.

35. The divergence between system and practice in mid-fifteenth-century local administration is the theme of Nimick's *Local Administration in Ming China,* ch. 2.

36. Gui Youguang, *Zhenchuan xiansheng ji,* 922–923.

37. *Changxing xianzhi* (1805), 7.3a–b.

38. Shen Defu, *Wanli yehuo bian,* 481. Shen makes this comment as a preface to the story of Suzhou silversmith Guan Fangzhou, whose story opens Chapter 9.

39. On Zhang Juzheng as a politican, see Huang, *1587, a Year of No Significance,* chs. 1–3.

40. Hsiao, "Mid-Yüan Politics," 552, 575, 585.

41. Wang Linheng, *Yuejian bian,* 92.

42. Von Glahn, *Fountain of Fortune,* provides a thorough review of monetary issues in the Ming, notably the shifting value of silver versus copper (157–160) and the debasement of copper coinage through the period (187–197).

43. Kuroda, "Copper Coins Chosen and Silver Differentiated," 67–74, has noted a significant divergence between what he calls "standard coins" (coins minted in the Song or early Ming, which people tended to save rather than spend) and "current coins" (later-minted coins of lower quality used for everyday buying and selling), the latter circulating at a large discount. A Hongwu "standard coin" appears on the cover of this book.

44. Huang Zongxi, *Huang Zongxi quanji,* vol. 2, 220.

45. Li Le, *Jianwen zaji,* 7.4a.

46. Yu Jideng, *Diangu jiwen,* 289. On the history of Ming granaries, see Hoshi Ayao, *Chūgoku no shakai fukushi no rekishi,* 55–81.

47. Yu Wenlong, *Shi luan,* 25.45b; Song Lian, *Yuan shi,* 4004.

48. Yu Wenlong, *Shi luan,* 25.59a–b; Song Lian, *Yuan shi,* 4332.

49. Even a philosopher as sober as Zhang Mou, who once declared that he wrote little because there was nothing to add to what the philosophers of the Song dynasty had already written, was prepared to abandon the dynastic founder's granary ideal; Goodrich and Fang, *Dictionary of Ming Biography,* 97.

50. Will and Wong, *Nourish the People,* 11–13.

51. Lu Zengyu, *Kangji lu,* 3a.48a–b.

52. *Lanyang xianzhi* (1545), 3.16a.

53. Gu Qing, "Su bian" (Changes in customs), reprinted in *Songjiang fuzhi* (1630), 7.23a-32a. For Gu's biography, see idem., 39.27b-29a; also Zhang Tingyu, *Ming shi,* 2432.

54. Girard, *Le Voyage en Chine d'Adriano de las Cortes,* 239. Elsewhere (165)

Las Cortes notes that "the Chinese eat very poorly," though that judgment may reflect the expectations of someone used to a meat-based diet. On Las Cortes, see Brook, *Vermeer's Hat,* 87–113. Pomeranz, *The Great Divergence,* 127–152, argues that the standard of living was higher in China than in Europe in the eighteenth century, and the polarization between rich and poor no greater.

55. Scott, *The Art of Not Being Governed,* 12–13.

56. Zhang Dai, *Taoan mengyi,* 110. Zhang's life is engagingly narrated in Spence, *Return to Dragon Mountain;* the hunt appears on p. 30.

57. Bray, *Technology and Society in Ming China,* 2–3.

58. Lau, *Mencius,* 164–165.

59. Menzies, "Forestry," 658–662. The state did undertake some reforestation along the northern border, though this was seen as a defense measure, not as resource renewal; see Qiu Zhonglin, "Mingdai changcheng yanxian de zhimu zaolin."

60. *Ming shizong shilu,* 202.2b.

61. Tan Qian, *Zaolin zazu,* 426, 453.

62. Elvin, *The Retreat of the Elephants,* 85.

63. Marks, *Tigers, Rice, Silk, and Silt,* 43.

64. Zhang, *Ming shi,* 5134. Qiao served as minister of personnel in the 1520s; the story probably dates to the turn of the sixteenth century.

65. *Qimen xianzhi* (1873), 36.4a–b, 6a–7a.

66. Jiao Hong, *Yutang congyu,* 266.

67. For another sighting south of the Yangzi about this time, see Sang Qiao, *Lushan jishi,* 1.39a, referring to an incident in 1551.

68. Yü, *The Renewal of Buddhism in China,* 20.

69. The last recorded sighting of a tiger in Guangdong and Guangxi was 1815; Marks, *Tigers, Rice, Silk, and Silt,* 325.

70. E.g., the demand for the pelts of sika deer on Taiwan was so great that they had almost disappeared by the middle of the seventeenth century; see Andrade, *How Taiwan Became Chinese,* 134–138, 149–150.

71. Zhang Dai, *Taoan mengyi,* 8.

6. Families

1. None but the emperor could wear images of the sun and moon; *Ming huidian,* 62.1a. The same ban also includes wearing dragons, phoenixes, lions, unicorns, or elephants. The imperial shoulder patches are noted in Li and Knight, *Power and Glory,* 259. Li and Knight include in their exhibition catalogue a gallery of official portraits of the Ming emperors that seems to indicate that these shoulder patches were not worn before the mid-fifteenth century (264).

2. Lu Rong, *Shuyuan zaji,* 62. Lu localizes this custom to Shanxi province, but it was probably practiced more widely.

3. Ye Chunji, *Huian zhengshu,* 4.6b.

4. Hazelton, "Patrilines and the Development of Localized Lineages."

5. Bray, *Technology and Gender,* 175–181.

6. For illustrations of these gender divisions, Sung Ying-hsing, *Chinese Technology in the Seventeenth Century,* 46, 101.

7. Farmer, *Zhu Yuanzhang and Early Ming Legislation,* 161; Birge, "Women and Confucianism from Song to Ming."

8. Cao Duan, "Jiagui jilüe" (Summary of family regulations), in his *Cao Yuechuan xiansheng ji* (Collected writings of Master Cao Yuechuan), quoted in Taga, *Chūgoku sōfu no kenkyū,* 168.

9. *Hejian fuzhi* (1540), 7.4b.

10. Dardess, *A Ming Society,* 97, 122–123.

11. *Yangzhou fuzhi* (1733), 34.11b.

12. E.g., *Hanyang fuzhi* (1546), 8.5b.

13. The following observations are derived from chaste widows' biographies in nine gazetteers: *Baoding fuzhi* (1607), *Daming fuzhi* (Zhengde), *Fengxiang fuzhi* (1766), *Funing zhouzhi* (1593), *Lianzhou fuzhi* (1637), *Nanchang fuzhi* (1588), *Qingzhou fuzhi* (1565), *Qiongzhou fuzhi* (1618), and *Yanzhou fuzhi* (1613). Chinese ages in lunar years (*sui,* which includes the year in which one is born) have been converted to Western ages.

14. Taga, *Chūgoku sōfu no kenkyū,* 169.

15. Franke, "Women under the Dynasties of Conquest," 41. The Ming Code followed this limitation; Jiang, *The Great Ming Code,* 214.

16. *Huian xianzhi* (1530), 9.6b–7a.

17. Dardess, *A Ming Society,* 81, using data from Jiangxi province; Liu Cuirong, *Ming-Qing shiqi jiazu renkou yu shehui jingji bianqian,* 97.

18. Li Rihua, *Weishui xuan riji,* 113. On theories of obstetrics and women's medicine in the Ming, see Furth, *A Flourishing Yin,* chs. 4–5.

19. Li Rihua, *Weishui xuan riji,* 173.

20. Chaoying Fang, "Huo T'ao," in Goodrich and Fang, *Dictionary of Ming Biography,* 681.

21. Zhao Jishi, *Jiyuan ji suoji,* vol. 1, 30.

22. The threat of divorce is exemplified in a story in Lu Rong, *Shuyuan zaji,* 47–48.

23. The novel has been recently retranslated by David Roy.

24. Shen Defu, *Wanli yehuo bian,* 459–460, tells an intricate story from the Hongzhi era of a woman named Mancang'er sold by her father into prostitution. After his death, her mother and brother tracked her down and tried to buy Mancang'er back, but she refused. The story turns on bribes and pay-offs, and ends with Hongzhi confiscating Mancang'er and sending her to serve in the Imperial Laundry.

25. Chang, *The Late Ming Poet Chen Tzu-lung.*

26. Brook, *The Confusions of Pleasure,* 97–99.

27. Lu Rong, *Shuyuan zaji,* 141–142; cf. Jiang, *The Great Ming Code,* 215.

28. Shen Defu, *Wanli yehuo bian,* 902; Xie Zhaozhe, *Wu zazu,* 8.4b; Szonyi, "The Cult of Hu Tianbao."

29. Liu Cuirong, *Ming-Qing shiqi jiazu renkou yu shehui jingji bianqian,* 53–55.

30. *Huian xianzhi* (1530), 9.6b.

31. Tan Qian, *Zaolin zazu,* 5; Chinese ages in *sui* have been converted to Western ages.

32. Elliott, "Hushuo: The Northern Other and *Han* Ethnogenesis."

33. Huang Qinglian, *Yuandai huji zhidu yanjiu,* 197–216. At the end of his master list, Huang notes that he suspects some categories have escaped his eye, and that "some day I will have to supplement these."

34. Ge Yinliang, *Jinling fancha zhi,* 1.33b. Zhu did not regard the four categories as exhausting the entire social spectrum, for he goes on to note the existence of Buddhist monks and Daoist priests as two categories of "experts" distinct from these four.

35. On the world of soldiering in the Ming, see Clunas, *Empire of Great Brightness,* 160–182.

36. Lu Rong, *Shuyuan zaji,* 134. The limitations on sons were waived under Yongle in the case of soldiers who had served in his campaign to take the throne.

37. Chen Wenshi, "Mingdai weisuo de junhu," 228.

38. Elman, *A Cultural History of Civil Examinations,* 140–143, 178.

39. Zhang Tingyu, *Ming shi,* 3336–3379.

40. On Shang Lu, see Zhang Tingyu, *Ming shi,* 4687–4691, and Tilemann Grimm's biography in Goodrich and Fang, *Dictionary of Ming Biography,* 1161–1163.

41. Li Le, *Jianwen zaji,* 1.43a. Though numbers were few, turnover was high; see Parsons, "The Ming Dynasty Bureaucracy."

42. *Ming shizong shilu,* 78.6a; also Coblin, "Brief History of Mandarin," 542.

43. The grand secretaries were Yang Rong (1371–1440) and Chen Shan (1365–1434). See Zhang Tingyu, *Ming shi,* 5741; Goodrich and Fang, *Dictionary of Ming Biography,* 1569.

44. Shinno, "Medical Schools and the Temples of the Three Progenitors in Yuan China"; Furth, *A Flourishing Yin,* 156–157.

45. Wang Daokun, *Taihan ji,* 492–493. Wang was a scholar in his own right, but also a writer for hire, "the best that money could buy"; Clunas, *Superfluous Things,* 14.

46. Brook, "Xu Guangqi in His Context," 80.

47. *Huzhou fuzhi* (1874), 44.10a.

48. Lu Rong, *Shuyuan zaji,* 85–86.

49. Brook, "Funerary Ritual and the Building of Lineages," 480. *The Family Rituals* has been translated by Patricia Ebrey.

50. This incident has been adapted from Brook, *The Chinese State in Ming Society*, 1–9.

51. *Jing'an xianzhi* (1565), 1.18a.

52. *Ming xiaozong shilu*, 155.4b–5a.

53. "Placard of the People's Instructions," in Farmer, *Zhu Yuanzhang and Early Ming Legislation*, 203.

54. Zhao Bingzhong, *Jiangxi yudi tushuo*, 2b.

55. Regarding eunuch grand defenders *(zhenshou)*, see Tsai, *Eunuchs in the Ming Dynasty*, 59–63.

56. See Zhang Tingyu, *Ming shi*, 5351, regarding an impeachment memorial against Dong, for which the Hongzhi emperor punished the official who submitted the memorial; also p. 4848, regarding an unsuccessful attempt to have the Zhengde emperor punish Dong.

57. Zhongyang tushuguan, *Mingren zhuanji ziliao suoyin*, 944; Jiao Hong, *Guochao xianzheng lu*, 90.9a.

58. *Ming xiaozong shilu*, 145.9b.

7. Beliefs

1. Liu and Berling, "The 'Three Teachings' in the Mongol-Yüan Period."

2. Brook, "Rethinking Syncretism."

3. Xie Zhaozhe, *Wu zazu*, 95.

4. Ge Yinliang, *Jinling fancha zhi*, 3.23a–26b, 64b; on Yuan imperial patronage, 1.17b.

5. Berger, "Miracles in Nanjing," 161.

6. Ge Yinliang, *Jinling fancha zhi*, 3.5a–7a.

7. *Analects* 3.12, translated in Legge, *The Confucian Classics*, vol. 1, 159.

8. Dean and Zheng, *Zongjiao beiming huibian: Quanzhou fu fence*, 961.

9. Tan Qian, *Zaolin zazu*, 222.

10. Brook, *The Chinese State in Ming Society*, 141–146.

11. Yu Jideng, *Diangu jiwen*, 107–108; Zhang Tingyu, *Ming shi*, 97.

12. This statement comes from the 1587 edition of the *Statutory Precedents of the Ming Dynasty*, the main compilation of imperial legislation: *Da Ming huidian*, 104.2a–b.

13. The sources for this and the following quotations from county gazetteers may be found in Brook, *The Chinese State in Ming Society*, 219–221: nn. 21 (Zhuozhou), 27 (Huairou), 52 (Linzhang), 73 (Nangong), and 74 (Qiuxian), in that order.

14. Brook, *Praying for Power*, 311–316.

15. On the problems of taking a quantitative approach to Chinese maps, see Yee, "Reinterpreting Traditional Chinese Geographical Maps," 53–67.

16. Zhang Huang, *Tushu bian,* 29.35a.

17. On reactions to the idea of a spherical earth, see Chu, "Trust, Instruments, and Cross-Cultural Scientific Exchanges"; also Yee, "Taking the World's Measure," 117–122. On Chinese responses to Jesuit cartography more generally, see Elman, *On Their Own Terms,* 122–131.

18. Zhang Huang, *Tushu bian,* 29. 33a, 39a.

19. Xu Guangqi, *Xu Guangqi ji,* 63.

20. Li Zhizao, preface to *Kunyu wanguo quantu,* in Li Tiangang, *Mingmo tianzhujiao sanzhushi wenjian zhu,* 148.

21. Li Zhizao, preface to *Hungai tongxian tushuo,* in Li Tiangang, *Mingmo tianzhujiao sanzhushi wenjian zhu,* 144.

22. Wang Qi, *Sancai tuhui, Dili* section, 1.1a.

23. Hashimoto, *Hsü Kuang-ch'i and Astronomical Reform,* 173, 189.

24. These letters appear in Li Zhi, *Fen shu,* 16–33, and Geng Dingxiang, *Geng Tiantai wenji,* 4.40a–45a. For a still influential assessment of Li Zhi, see Huang, *1587, a Year of No Significance,* 189–221.

25. *Analects* 12:1, translated in Legge, *The Confucian Classics,* vol. 1, 250.

26. *Analects* 17:13, translated in Legge, *The Confucian Classics,* vol. 1, 324.

27. Letter by Ma Jinglun reprinted in Xiamen daxue, *Li Zhi yanjiu cankao ziliao,* 64.

28. Cai Ruxian, *Dongyi tuxiang, zongshuo,* 2a.

29. *Chaoyi xianzhi* (1519; 1824), *fengsu,* 9a; the same sentiment is expressed in *Suiyao tingzhi* (1873), *fengsu,* 18b.

30. Quoted in Scott, *The Art of Not Being Governed,* 13.

31. Bol, *Neo-Confucianism in History,* 216.

32. Wang Zhichun, *Chuanshan gong nianpu,* 1.20b–21a.

33. Li Zhizao, preface to *Kunyu wanguo quantu,* in Li Tiangang, *Mingmo tianzhujiao sanzhushi wenjian zhu,* 149.

34. Xu Guangqi, "Zhengdao tigang" (Outline of the true way), in Li Tiangang, *Mingmo tianzhujiao sanzhushi wenjian zhu,* 107. My thanks to Li Tiangang for drawing my attention to this feature of Xu's thought.

35. Yang Tingyun, *Daiyi xubian* (Sequel to "Treatise to Supplant Doubts"), quoted in Standaert, *Yang Tingyun,* 206–208, with slight minor revisions to the translation.

8. The Business of Things

1. Clunas, *Superfluous Things,* 46. Clunas provides a summary of the inventory on pp. 47–48. On Yan Song, see the biography by Kwan-wai So in Goodrich and Fang, *Dictionary of Ming Biography,* 1586–1591.

2. These inventories have been recovered by the historian Wu Renshu, who presents them in his book, *Pinwei shehua,* 225–232.

3. Dudink, "Christianity in Late Ming China," 177–226.

4. Cao's book is translated in full in David, *Chinese Connoisseurship.*

5. Weitz, *Zhou Mi's Record of Clouds and Mist,* 4, 20.

6. E.g., *Changshu xianzhi* (1539), 4.20b.

7. Brook, *The Confusions of Pleasure,* 144–147.

8. On the transition from corvée obligation among carpenters, see Ruitenbeek, *Carpentry and Building in Late Imperial China,* 16–17.

9. Jiangsu sheng bowuguan, *Jiangsu sheng Ming Qing yilai beike ziliao xuanji,* 135–136.

10. Ming court art is examined in Barnhart, *Painters of the Great Ming.*

11. Chaoying Fang has a biography of Li Rihua in Goodrich and Fang, *Dictionary of Ming Biography,* vol. 1, 826–830. Li's social context is explored in Li Chu-tsing, "Li Rihua and his Literati Circle in the Late Ming Dynasty." For two examples of his painting and one of his calligraphy, see Li Chu-tsing, *The Chinese Scholar's Studio,* plates 3, 4c, and 5. See Barnhart, *The Jade Studio,* 116–117, for further samples of his calligraphy.

12. *Hengzhou fuzhi* (1536), 9.14b.

13. The techniques of book production are sketched in McDermott, *A Social History of the Chinese Book,* 9–42. On the mechanics of the printed page, see Chia, *Printing for Profit,* 25–62.

14. Gu Yanwu, *Gu Tinglin shiwen ji,* 29–30. Gu reports that the collection was scattered and lost following the Manchu invasion.

15. Brook, *The Chinese State in Ming Society,* 101.

16. Wu Han, *Jiang Zhe cangshujia shilüe,* 10.

17. The references to books in Li's diary are taken from Li Rihua, *Weishui xuan riji,* 73, 105, 190–191, 277–278, 303, 305, 374, 454–455, 496. On the high valuation of Song imprints, see Dong Qichang, *Yunxuan qingbi lu,* 21–22.

18. Unschuld, *Medicine in China: A History of Pharmaceutics,* 128–142.

19. Brook, *The Chinese State in Ming Society,* 128–129.

20. Girard, *Le Voyage en Chine d'Adriano de las Cortes,* 191, 193. On early childhood education in the Ming, see Schneewind, *Community Schools and the State in Ming China.*

21. On the publishing trade, see Chow, *Printing, Culture, and Power,* 57–89.

22. On Ming novels, see Plaks, *The Four Masterworks of the Ming Novel.* These novels have been translated by Pearl Buck, Arthur Waley, and David Roy.

23. Tang Shunzhi, *Jingchuan xiansheng youbian.*

24. The Jianyang publishing industry is explored in Chia, *Printing for Profit.*

25. Translated by Thomas Cleary as *The Flower Ornament Scripture.*

26. Clunas, *Chinese Furniture,* 19.

27. Fan Lian, *Yunjian jumu chao,* ch. 2, adapted from the translation in Ruitenbeek, *Carpentry and Building,* 15.

28. Quoted in Clunas, *Superfluous Things,* 42. This handbook is discussed in the next chapter.

29. Clunas, *Superfluous Things,* 145, citing Zhan Han's "On Artisans" (*Baigong ji*).

30. Girard, *Le Voyage en Chine d'Adriano de las Cortes,* 250.

31. Clunas, *Chinese Furniture,* 55.

32. Wu Renshu, *Pinwei shehua,* 228–229.

33. Clunas, *Superfluous Things,* 63.

34. For Li's furniture references, see *Weishui xuan riji,* 164, 246, 481.

35. Watt and Leidy, *Defining Yongle,* 27–30.

36. Carswell, *Blue and White,* 17.

37. Li Rihua, *Weishui xuan riji,* 92.

38. For an amusing tussle between Li and his dealer, see Brook, *Vermeer's Hat,* 80–81.

39. Weitz, *Zhou Mi's Record of Clouds and Mist,* 238–239.

40. For Li's references to paintings, see *Weishui xuan riji,* 58, 62, 93, 124, 170, 187, 283, 298, 417.

41. On Wen Zhengming's "coarse" style, see Clunas, *Elegant Debts,* 178.

42. Cahill, *Parting at the Shore,* 9–14.

43. The term is taken from Clunas, *Elegant Debts,* 8.

44. Revised from the translation in Clunas, *Elegant Debts,* 176.

45. Li Rihua, *Weishui xuan riji,* 406.

46. Hsü, *A Bushel of Pearls,* 16.

9. The South China Sea

1. Jiang, *The Great Ming Code,* 157–158, 244. On the perceived difference between strangulation and decapitation, see Brook et al., *Death by a Thousand Cuts,* 50–51. Ming law accepted the ideal that the corpse should be preserved whole, and allowed punishments to transgress that ideal only when a criminal committed the very worst crimes.

2. Shen Defu, *Wanli yehuo bian,* 481.

3. Bodley, *The Life of Sir Thomas Bodley,* 38, 58.

4. Trevor-Roper, *Archbishop Laud,* 276.

5. The rutter was transcribed and printed in 1961 by Xiang Da in *Liangzhong haidao zhenjing.* For its connections to texts from the Zheng He expeditions, see Tian Rukang, "*Duhai fangcheng.*"

6. Rosenblatt, *Renaissance England's Chief Rabbi. De diis Syriis* (On the gods of the Syrians) launched his reputation as an Orientalist when it was pub-

lished in 1617; a series of studies of Hebraic law sustained that reputation through the 1630s.

7. On Selden's relationship with Laud, see Trevor-Roper, *Archbishop Laud,* 336–337.

8. The map has remained unknown for the simple reason that no one has asked to see it for decades. Even Li Xiaocong, the authority on Chinese maps abroad, missed it when he visited Oxford in 1992 while compiling his *Descriptive Catalogue of Pre-1900 Chinese Maps Seen in Europe.* I am immensely grateful to David Helliwell for bringing the map to my attention and making a copy of it available to me.

9. So, *Prosperity, Regions, and Institutions in Maritime China,* 117–125.

10. Ye Sheng, *Shuidong riji,* 17.2a.

11. Ledyard, "Cartography in Korea," 244–246.

12. Zhang Tingyu, *Ming shi,* 23–28, 34–35; *Ming taizu shilu,* 18 Sept. 1397, translated by Geoff Wade in *Southeast Asia in the Ming Shi-lu.*

13. Zhang Tingyu, *Ming shi,* 717–776, 80.

14. Publisher Mao Yuanyi (1594–ca. 1641) includes pictorial maps of Zheng's route in his massive 1621 compendium, *Wubei zhi (Records of Military Preparedness).*

15. Xie Zhaozhe, *Wu zazu (Five Miscellanies),* 272, 360–361; the latter entry is translated in Elvin, *The Retreat of the Elephants,* 378–379, which I have revised in minor ways. Tan Qian also reports the incident in his *Zaolin zazu,* 483, probably on the basis of Xie's account.

16. Jianwen's order, issued in 1401, gave those holding stocks of foreign merchandise three months to sell them off; *Guangdong tongzhi* (1822), 187.6.

17. *Ming xiaozong shilu,* 73.3a–b; Zhang Tingyu, *Ming shi,* 4867–4868.

18. Zhang Tingyu, *Ming shi,* 212.

19. These developments are treated further in Brook, *The Confusions of Pleasure,* 119–124.

20. *Quanzhou fuzhi* (1829), 73.20a–32a.

21. Tan Qian, *Zaolin zazu,* 571, 580.

22. Brook, *Vermeer's Hat,* 100–107.

23. Memorial of Fu Yuanchu, copied into Gu Yanwu, *Tianxia junguo libing shu,* 26.33a.

24. Alves, "La voix de la prophétie," 41–44.

25. Braudel, *The Perspective of the World,* 21–22.

26. Zhang Xie, *Dongxi yang kao, fanli,* 20. These routes are recorded on 171–185.

27. Andrade, *How Taiwan Became Chinese,* 20.

28. Blussé, *Visible Cities,* 58–60, 64–65.

29. Gu Yanwu, *Tianxia junguo libing shu,* 26.33b.

30. The 1639–40 massacre is described in Brook, *Vermeer's Hat,* ch. 6.

31. On the cultural consumption of Ming gardens, see Clunas, *Fruitful Sites;* Clunas devotes an entire chapter to gardens owned by the Wen family (104–136).

32. Clunas, *Superfluous Things,* 41 (with slight modifications), 43.

33. Zhang Xie, *Dongxi yang kao:* Zhou Qiyuan's preface, 17.

34. Shen Defu, *Wanli yehuo bian,* 783.

35. Shen Defu, *Wanli yehuo bian,* 783.

36. Brockey, *Journey to the East,* 29–30.

37. Standaert, *Handbook of Christianity in China,* 291.

38. Shen Que, quoted in Brook, *Vermeer's Hat,* 108.

39. Peterson, "Why Did They Become Christians?"

40. Ricci is the subject of several outstanding biographies, notably Spence, *The Memory Palace of Matteo Ricci,* and Hsia, *A Jesuit in the Forbidden City.*

41. Shen Defu, *Wanli yehuo bian,* 783.

42. Menegon, *Ancestors, Virgins, and Friars.*

10. Collapse

1. Huang Zongxi, *Hongguang shilu chao,* in his *Huang Zongxi quanji,* vol. 2, 1. Huang did not sign the work, but the work is attributed to him and is regarded as expressing the views of Huang's circle; Struve, *The Ming-Qing Conflict,* 226.

2. Huang Zongxi, *Huang Zongxi quanji,* vol. 2, 3.

3. Li Qing, *Sanyuan biji,* 90.

4. Ye Mengzhu, *Yueshi bian,* 183.

5. Wang Wei, "Parting in the Boat on an Autumn Night," translated by Kang-i Sun Chang, in Chang and Saussy, *Women Writers of Traditional China,* 322.

6. The narrative of decline shapes the work of the two main histories of the Ming written in English by Chinese historians in the 1980s: Ray Huang's *1587, a Year of No Significance: The Ming Dynasty in Decline,* and Albert Chan's *The Glory and Fall of the Ming Dynasty.*

7. Huang, "The Lung-ch'ing and Wan-li Reigns," 517; idem., "The Ming Fiscal Administration," 162–164.

8. Yang Dongming, *Jimin tushuo.*

9. *Ming shenzong shilu,* 538.2b, 539.9b, 540.7b, 542.2b.

10. Zhang Tingyu, *Ming shi,* 512.

11. Huang, "The Lung-ch'ing and Wan-li Reigns," 583.

12. For an early study of soaring taxes and tax defaults, see Wang Yü-ch'üan, "The Rise of Land Tax and the Fall of Dynasties in Chinese History."

13. Huang Yi-Long, "Sun Yuanhua."

14. Brook, *Vermeer's Hat,* 103–104.

15. Huang Yi-Long, "Sun Yuanhua," 250–255.
16. *Ming xizong shilu,* 36.2b.
17. Li Qing, *Sanyuan biji,* 8.
18. Tan Qian, *Zaolin zazu,* 597–598.
19. Wakeman, *The Great Enterprise,* 130.
20. Li Qing, *Sanyuan biji,* 17.
21. Russian coldness data roughly coincides with Chinese grain price data. As well, Russian wetness/dryness data mildly confirms the annual findings in the Central Meteorological Bureau maps; see Lamb, *Climate,* vol. 2, 562, 564.
22. *Ming chongzhen changbian,* 57.6a, 63.10a, 64. 20b.
23. *Jinan fuzhi* (1840), 20.18b; *Deping xianzhi* (1673), 3.40a.
24. *Linqing zhouzhi* (1674), 3.40a.
25. *Shanghai xianzhi* (1882), 30.9b. The normal price of a peck of rice was one-thirtieth of an ounce.
26. Zhang Tingyu, *Ming shi,* 486.
27. Cooper, *Rodrigues the Interpreter,* 342, 346.
28. *Gansu xin tongzhi* (1909), 2.36a.
29. Zhang Tingyu, *Ming shi,* 477.
30. *Yizhou zhi* (1674), 1.8b.
31. *Caozhou zhi* (1674), 19.21a; *Xinzheng xianzhi* (1693), 4.96a.
32. Dunstan, "The Late Ming Epidemics"; Hanson, "Inventing a Tradition," 103–107.
33. *Yunzhong zhi* (1652), 12.20a; the epidemic faded away later that year.
34. *Ming xizong shilu,* 33.15a.
35. Li Qing, *Sanyuan biji,* 3. On the local impact of the financial crisis, see Nimick, *Local Administration.*
36. Parsons, *Peasant Rebellions of the Late Ming Dynasty,* 4–6; see pp. 17–21 for portraits of Li Zicheng and Zhang Xianzhong. Maps showing the locations of rebellion for each year from 1628 to 1642 are distributed across pp. 3–84.
37. On Li Zicheng's campaigns through the early 1640s, see Des Forges, *Cultural Centrality and Political Change in Chinese History,* 204–311.
38. Zhao Shiyu and Du Zhengzhen, "'Birthday of the Sun.'"
39. *Xinzheng xianzhi* (1693), 4.100a.
40. This history is reconstructed in Struve, *The Southern Ming.*
41. Huang Zongxi, *Huang Zongxi quanji,* vol. 2, 205–206.
42. For a selection of such stories in translation, see Struve, *Voices from the Ming-Qing Cataclysm.*
43. Huang Zongxi, *Huang Zongxi quanji,* vol. 2, 240.
44. Ibid., 239.
45. Cahill, *The Compelling Image,* 186–225.
46. Huang Yuanjie, "Qingming Festival, 1646," modified from the translation

by Kang-i Sun Chang in Chang and Saussy, *Women Writers of Traditional China,* 359.

47. Shang Jinglan, "Mourning the Dead: In Memory of my Husband," modified from the translation by Ellen Widmer in Chang and Saussy, *Women Writers of Traditional China,* 320.

Conclusion

1. Abu-Lughod, *Before European Hegemony,* p. 12.

2. Bray, *Agriculture,* 489–490; Brook, *The Chinese State in Ming Society,* 85–89.

3. Wang Zhen, *Wang Zhen nongshu,* 6–9. Regarding other such texts, see Zhou Zhiyuan, *Mingdai huangzheng wenxian yanjiu,* 33–59. The first famine pharmacopoeia, *Jiuhuang bencao* (Materia media for survival during famines), was compiled by Zhu Yuanzhang's fifth son and published in 1406; Unschuld, *Medicine in China,* 221.

4. I wish to thank James Wilkerson for helping me clarify the logic of this argument.

5. Brook, *Vermeer's Hat,* 222–223.

6. Zhang Xie, *Dongxi yang kao,* 170.

7. "Climbing the Stork Pagoda (after Wang Zin-Huai)," from Ron Butlin, *The Exquisite Instrument,* 29; cited with the kind permission of the author.

8. Cahill, *The Compelling Image,* 83.

9. *Shanxi tongzhi* (1682), 30.41b.

BIBLIOGRAPHY

For the convenience of readers, the bibliography is divided between primary sources in Chinese and secondary sources in all languages, primarily English. The local gazetteers cited in the endnotes do not appear as entries in the bibliography.

Primary Sources

Cai Ruxian. *Dongyi tuxiang* (Illustrations of foreigners from the Eastern Sea). 1586.

Chen Jian. *Huang Ming congxin lu* (Record of official activists of the Ming dynasty), ed. Shen Guoyuan. 1620.

Chen Quanzhi. *Pengchuang rilu* (Daily notes from Rattan Window). Preface dated 1565. Shanghai, 1979.

Chen Yaowen. *Tianzhong ji* (All within Heaven). Preface dated 1569. Reprinted 1589.

Chen Zilong, ed. *Ming jingshi wenbian* (Collected statecraft writings of the Ming). Huating, 1638. Reprint, Beijing: Zhonghua shuju, 1982.

Chongzhen changbian. See *Ming shilu*.

Da Yuan shengzheng guochao dianzhang (Dynastic institutions of the sagely administration of the Great Yuan). 1303. Photoreprint, Taipei: National Palace Museum, 1972.

Dean, Kenneth, and Zheng Zhenman, eds. *Fujian zongjiao beiming hiubian: Quanzhou fu fence* [Epigraphic materials on the history of religion in Fujian: Quanzhou region]. 3 vols. Fuzhou: Fujian renmin chubanshe, 2003.

Dong Qichang. *Yunxuan qingbi lu* (Pure jottings from Yun Pavilion). Reprinted with Chen Jiru, *Nigu lu* (Record of my fondness for antiquities). 1937.

Fuzheng, ed. *Hanshan dashi nianpu shuzhu* (Annotated chronological biography of Master Hanshan). Reprint, Taipei: Zhenshanmei chubanshe, 1967.

Ge Yinliang. *Jinling fancha zhi* (Gazetteer of the Buddhist monasteries of Nanjing). Nanjing: Senglusi, 1607. Reprint, 1627.

Geng Dingxiang. *Geng Tiantai wenji* (Collected works of Geng Tiantai).

Gu Qiyuan. *Kezuo zhuiyu* (Superfluous chats from the guest's seat). Nanjing, 1617.

Gu Yanwu. *Gu Tinglin shiwen ji* (Collected poetry and prose writings of Gu Tinglin). Beijing: Zhonghua shuju, 1983.

———. *Tianxia junguo libing shu* (The strengths and weaknesses of the various regions of the realm). 1662. Reprint, Shanghai: Shanghai guji chubanshe, 1984.

Gui Youguang. *Zhenchuan xiansheng ji* (The collected writings of Master Zhenchuan). Reprint, Shanghai: Shanghai guji chubanshe, 1981.

Hai Rui. *Hai Rui ji* (Collected writings of Hai Rui). Reprint, Beijing: Zhonghua shuju, 1981.

Huai Xiaofeng, ed. *Da Ming lü* (Ming Code). Beijing: Falü chubanshe, 1999.

Huang Yu. *Shuanghuai suichao* (Notes by the year from the Master of the Double Locust Trees). Compiled 1456–97, published 1549. Reprint, Beijing: Zhonghua shuju, 1999.

Huang Zongxi. *Huang Zongxi quanji* (Complete works of Huang Zongxi). Hangzhou: Zhejiang guji chubanshe, 1985.

Huanyu tongqu (Network of routes connecting the realm). Nanjing: Bingbu (Ministry of war), 1394. Photoreprinted in Siku quanshu cunmu congshu (1997), vol. 166.

Jiangsu sheng bowuguan (Jiangsu Provincial Museum), ed. *Jiangsu sheng Ming Qing yilai beike ziliao xuanji* (Selected epigraphic materials since the Ming and Qing from Jiangsu province). Beijing: Sanlian shudian, 1959.

Jiao Hong. *Guochao xianzheng lu* (Biographies of accomplished administrators of this dynasty). 1616.

———. *Yutang congyu* (Comments from Jade Hall). 1618. Reprint, Beijing: Zhonghua shuju, 1981.

Lang Ying. *Qixiu leigao* (Categorized drafts revised seven times). Reprint, Taipei: Shijie shuju, 1963.

Lei Menglin. *Dulü suoyan* (Sundry comments on reading the Code). 1557. Reprint, Beijing: Falü chubanshe, 2000.

Li Le. *Jianwen zaji* (Random notes on what I have seen and heard). Postface dated 1610. Reprint, Shanghai: Shanghai guji chubanshe, 1986.

Li Qing. *Sanyuan biji* (Notes from the three-walled ministry [of justice]). Compiled 1640s. Beijing: Zhonghua shuju, 1982.

Li Tiangang, ed. *Mingmo tianzhujiao sanzhushi wenjian zhu: Xu Guangqi, Li Zhizao, Yang Tingyun lunjiao wenji* (Annotated writings of the three founders of Catholicism in the late Ming: Christian documents of Xu Guangqi, Li Zhizao, and Yang Tingyun). Hong Kong: Daofeng shushe, 2007.

Li Zhi. *Fen shu, Xu fen shu* (A book to be burned; A further book to be burned). Reprint, Beijing: Zhonghua shuju, 1975.

Lin Zhaoke. *Kaogong ji shuzhu* (Annotated record of the scrutiny of craftsmen). Jianyang, 1603.

Liu Ji, ed. *Da Ming qinglei tianwen fenye zhi shu* (Book of the earthly correspondences of the heavenly constellations under the Great Ming). 1384.

Lu Can. *Gengsi bian* (Notes from the last two years of the Zhengde reign). Compiled 1510–19, first published 1590. Reprint, Beijing: Zhonghua shuju, 1987.

Lu Rong. *Shuyuan zaji* (Miscellany from Bean Garden). Compiled before 1494. Reprint, Beijing: Zhonghua shuju, 1987.

Lu Zengyu. *Kangji lu* (Record of aids to prosperity), ed. Ni Goulian. 1740.

Ming huidian (Collected statutes of the Ming), ed. Shen Shixing. 1588.

Ming shilu (Veritable records of the Ming dynasty): *Ming taizu shilu* (Veritable records of Emperor Taizu [Hongwu, 1368–98]); *Ming xiaozong shilu* (Veritable records of Emperor Xiaozong [Hongzhi, 1488–1505]); *Ming wuzong shilu* (Veritable records of Emperor Wuzong [Zhengde, 1506–21]); *Ming shizong shilu* (Veritable records of Emperor Shizong [Jiajing, 1522–66]); *Ming shenzong shilu* (Veritable records of Emperor Shenzong [Wanli, 1573–1620]); *Ming xizong shilu* (Veritable records of Emperor Xizong [Tianqi, 1621–27]); *Chongzhen changbian* (Unedited compilation for the Chongzhen reign). Taipei: Zhongyang yanjiuyuan lishi yuyan yanjiusuo, 1962.

Sang Qiao. *Lushan jishi* (A record of events pertaining to the Lu Mountains). 1561.

Shen Bang. *Wanshu zaji* (Miscellaneous records from the Wanping county office). 1593. Reprint, Beijing: Beijing guji chubanshe, 1980.

Shen Defu. *Wanli yehuo bian* (Unofficial gleanings from the Wanli era). Preface dated 1619. Reprint, Beijing: Zhonghuo shuju, 1997.

Shen Jiaben. *Shen Jiyi xiansheng yishu yibian* (Surviving writings of Shen Jiyi [Jiaben], second collection). Reprint, Taipei: Wenhai chubanshe, 1967.

Song Lian, ed. *Yuan shi* (Official history of the Yuan dynasty). Completed 1371. Reprint, Beijing: Zhonghua shuju, 1976.

Song Yingxing. *Tiangong kaiwu* (The creation of things by heaven and artifice). See Sung Ying-hsing, *T'ien-kung k'ai-wu.*

Tan Qian. *Zaolin zazu* (Date grove miscellany). Preface dated 1644; first published 1911. Reprint, Beijing: Zhonghua shuju, 2006.

Tang Shunzhi. *Jingchuan xiansheng youbian* (Master Jingchuan's compendium of administrative writings). Nanjing: Guozijian, 1595. Reprinted in the Siku quanshu cunmu congshu, series 2, vols. 70–71. Jinan: Qi Lu shushe, 1997.

Tao Chengqing. *Da Ming yitong wenwu zhusi yamen guanzhi* (The administrative system for civil and military officials throughout the unified realm of the Great Ming), ed. Ye Shiyong. Jiangxi, 1586.

Tao Zongyi. *Nancun chuogeng lu* (Master Nancun's notes after the plowing is done). First published 1366. Reprint, Beijing: Zhonghua shuju, 2004.

Wang Daokun. *Taihan ji* (The collected writings of Master Taihan). 4 vols. Reprint, Hefei: Huangshan shushe, 2004.

Wang Linheng. *Yuejian bian* (The Guangdong sword compilation). Reprinted with Ye Quan, *Xianbo bian,* and Li Zhongfu, *Yuanli erzai.* Beijing: Zhonghua shuju, 1987.

Wang Qi. *Sancai tuhui* (Illustrated congress of the three powers). 1607.

Wang Shixing. *Guangzhi yi* (Further record of my extensive travels). Completed 1597, first published 1644. Reprint, Beijing: Zhonghua shuju, 1981.

Wang Yangming [Shouren]. *Wang Yangming quanji* (The complete works of Wang Yangming). Reprint, Shanghai: Shanghai guji chubanshe, 1992.

Wang Zhen. *Wang Zhen nongshu* (Wang Zhen's agricultural manual). First published 1313. Bejing: Nongye chubanshe, 1981.

Wang Zhichun. *Chuanshan gong nianpu* (Annual biography of Master Chuanshan [Wang Fuzhi]). 1893. Reprint, Hengyang: Hengyang shi bowuguan, 1975.

Xiang Da, ed. *Liangzhong haidao zhenjing* (Two maritime navigation manuals). Beijing: Zhonghua shuju, 1961.

Xiao Xun. *Gugong yilu* (Record of the remnants of the Old Palace). Written 1368, preface dated 1398. Reprint, Beijing, 1616.

Xie Zhaozhe. *Wu zazu* (Five offerings). Late Wanli era (1610s). Reprint, Shanghai: Shanghai shudian, 2001.

Xu Guangqi. *Nongzheng quanshu* (Complete handbook of agricultural administration). Reprint, Shanghai: Shanghai guji chubanshe, 1979.

———. *Xu Guangqi ji* (The collected works of Xu Guangqi). 2 vols. Shanghai: Shanghai guji chubanshe, 1984.

Xu Hongzu. *Xu Xiake youji* (Travel diaries of Xu Xiake), ed. Ding Wenjiang. 1928. Reprint, Beijing: Shangwu yinshuguan, 1996.

Yang Dongming. *Jimin tushuo* (An album of the famished). 1658, reprinted 1748.

Yang Zhengtai, ed. *Tianxia shuilu lucheng* (Water and land routes of the realm). Taiyuan: Shanxi renmin chubanshe, 1992.

Yao Yu. *Linghai yutu* (Terrestrial map of maritime Guangdong). 1542. Photoreprint of the Siku quanshu edition, Taipei: Guangwen shuju, 1969.

Ye Chunji. *Huian zhengshu* (Administrative handbook of Huian county). 1573. Reprinted in his *Shidong wenji* (Collected writings of Master Shidong). 1672.

Ye Mengzhu. *Yueshi bian* (A survey of the age). Kangxi era. Shanghai: Shanghai guji chubanshe, 1981.

Ye Sheng. *Shuidong riji* (Notes compiled daily from east of the river). Siku quanshu edition, 1778.

Ye Ziqi. *Caomuzi* (The scribbler). 1378, posthumously published in 1516. Reprint, Beijing: Zhonghua shuju, 1959.

Yu Jideng. *Diangu jiwen* (Notes on what I have heard regarding statutory precedents). First published Wanli era. Reprint, Beijing: Zhonghua shuju, 2006.

Yu Wenlong. *Shi luan* (Shreds of history). 1618.

Yu Xiangdou, ed. *Wanyong zhengzong* (The correct source for a myriad practical uses). Jianyang, 1599.

Yuan dianzhang. See *Da Yuan shengzheng guochao dianzhang*.

Zhang Huang, ed. *Tushu bian* (Illustrations and texts). 1613.

Zhang Tingyu, ed. *Ming shi* (Official history of the Ming dynasty). Reprint, Beijing: Zhonghua shuju, 1974.

Zhang Xie. *Dongxi yang kao* (Studies of the Eastern and Western Seas). Reprint, Beijing: Zhonghua shuju, 1981.

Zhang Yi. *Yuguang jianqi ji* (The jade-bright sword collection). 2 vols. Reprint, Beijing: Zhonghua shuju, 2006.

Zhao Bingzhong. *Jiangxi yudi tushuo* (Illustrated geography of Jiangxi). Wanli era.

Zhao Erxun, ed. *Qing shi gao* (Draft history of the Qing dynasty). Reprint, Beijing: Zhonghua shuju, 1976.

Zhao Jishi. *Jiyuan ji suoji* (Reliance garden relies on what it relies on). 1695.

Zhou Mi. *Yunyan guoyan lu* (Record of clouds and mist passing before one's eyes). 1296. See Weitz, Ankeney.

Zhu Yuanzhang. *Ming taizu ji* (Collected writings of the founding Ming emperor). First published 1374. Reprint, Hefei: Huangshan shushe, 1991.

———. *Yuzhi dagao* (Imperially compiled grand pronouncements); *Yuzhi xugao* (Imperially compiled grand pronouncements, further compilation); *Yuzhi sangao* (Imperially compiled grand pronouncements, third compilation). Reprinted in Yang Yifan, *Ming dagao yanjiu* (Studies in the grand pronouncements of the Ming). Nanjing: Jiangsu renmin chubanshe, 1988.

Secondary Sources

Abu-Lughod, Janet. *Before European Hegemony: The World-System A.D. 1250–1350*. New York: Oxford University Press, 1989.

Allsen, Thomas. "The Rise of the Mongolian Empire and Mongolian Rule in North China." In *The Cambridge History of China*, vol. 6: *Alien Regimes and Border States 907–1368*, ed. Herbert Franke and Denis Twitchett, 321–413. Cambridge: Cambridge University Press, 1994.

Alves, Jorge M. dos Santos. "La voix de la prophétie: Informations portuguaises de la 1e moitié du XVIe siècle sur les voyages de Zheng He." In *Zheng He:*

Images and Perceptions/Bilder und Wahrnehmingen, ed. Claudine Salmon and Roderich Ptak, 39–55. Wiesbaden: Harrassowitz, 2005.

Andersson, Gunnar. *Children of the Yellow Earth: Studies in Prehistoric China*. London: Kegan Paul, Trench, Trübner, 1973.

Andrade, Tonio. *How Taiwan Became Chinese: Dutch, Spanish, and Han Colonization in the Seventeenth Century*. New York: Columbia University Press, 2008.

Barnhart, Richard. *Painters of the Great Ming: The Imperial Court and the Zhe School*. Dallas: Dallas Museum of Art, 1993.

——— et al. *The Jade Studio: Masterpieces of Ming and Qing Painting and Calligraphy from the Wong Nan-p'ing Collection*. New Haven: Yale University Art Gallery, 1994.

Benedictow, Ole. *The Black Death, 1346–1353: The Complete History*. Woodbridge: Boydell Press, 2004.

Berger, Patricia. "Miracles in Nanjing: An Imperial Record of the Fifth Karmapa's Visit to the Chinese Capital." In *Cultural Intersections in Later Chinese Buddhism*, 145–169. Honolulu: University of Hawaii Press, 2001.

Birge, Bettine. "Women and Confucianism from Song to Ming: The Institutionalization of Patrilineality." In *The Song-Yuan-Ming Transition in Chinese History*, ed. Paul Jakov Smith and Richard von Glahn, 212–240. Cambridge, Mass.: Harvard University Press, 2003.

Blue, Gregory. "China and Western Social Thought in the Modern Period." In *China and Historical Capitalism: Genealogies of Sinological Knowledge*, ed. Timothy Brook and Gregory Blue, 57–109. Cambridge: Cambridge University Press, 1999.

Blussé, Leonard. *Visible Cities: Canton, Nagasaki, and Batavia and the Coming of the Americans*. Cambridge, Mass.: Harvard University Press, 2009.

Bodley, Thomas. *The Life of Sir Thomas Bodley*. Chicago: A. C. McClurg, 1906.

Bol, Peter. "Geography and Culture: Middle-Period Discourse on the *Zhong guo*—the Central Country." *Hanxue yanjiu*, 2009.

———. *Neo-Confucianism in History*. Cambridge, Mass.: Harvard University Asia Center, 2008.

Braudel, Fernand. *The Perspective of the World*. Vol. 3 of *Civilization and Capitalism, 15th–18th Century*. London: Collins, 1984.

Bray, Francesca. *Agriculture. Science and Civilisation in China*, VI:2, ed. Joseph Needham. Cambridge: Cambridge University Press, 1984.

———. *Technology and Gender: Fabrics of Power in Late Imperial China*. Berkeley: University of California Press, 1997.

———. *Technology and Society in Ming China (1368–1644)*. Washington, DC: American Historical Association, 2000.

Brockey, Liam. *Journey to the East: The Jesuit Mission to China, 1579-1724*. Cambridge, Mass.: Harvard University Press, 2007.

Brokaw, Cynthia, and Kai-wing Chow, eds. *Printing and Book Culture in Late Imperial China.* Berkeley: University of California Press, 2005.

Brook, Timothy. *The Chinese State in Ming Society.* London: RoutledgeCurzon, 2005.

———. "Communications and Commerce." In *The Cambridge History of China,* vol. 8: *The Ming Dynasty,* pt. 2, ed. Denis Twitchett and Frederick Mote, 579–707. Cambridge: Cambridge University Press, 1998.

———. *The Confusions of Pleasure: Commerce and Culture in Ming China.* Berkeley: University of California Press, 1998.

———. "The Early Jesuits and the Late Ming Border: The Chinese Search for Accommodation." In *Encounters and Dialogues: Changing Perspectives on Chinese-Western Exchanges from the Sixteenth to Eighteenth Centuries,* ed. Xiaoxin Wu, 19–38. Sankt Augustin: Monumenta Serica, 2005.

———. "Europaeology? On the Difficulty of Assembling a Knowledge of Europe in China." In *Christianity and Cultures: Japan and China in Comparison (1543–1644),* ed. Antoni Ucerler, 261–285. Rome: Institutum Historicum Societatis Iesu, 2010.

———. "Funerary Ritual and the Building of Lineages in Late Imperial China." *Harvard Journal of Asiatic Studies* 49, 2 (December 1989): 465–499.

———. *Geographical Sources of Ming-Qing History,* 2nd enlarged ed. Ann Arbor: Center for Chinese Studies, University of Michigan, 2002.

———. *Praying for Power: Buddhism and the Formation of Gentry Society in Late-Ming China.* Cambridge, Mass.: Council on East Asian Studies, Harvard University, 1993.

———. "Rethinking Syncretism: The Unity of the Three Teachings and their Joint Worship in Late-Imperial China." *Journal of Chinese Religions* 21 (Fall 1993): 13–44.

———. *Vermeer's Hat: The Seventeenth Century and the Dawn of the Global World.* New York: Bloomsbury; Toronto: Viking; London: Profile, 2008.

———. "What Happens When Wang Yangming Crosses the Border?" In *The Chinese State at the Borders,* ed. Diana Lary, 74–90. Vancouver: University of British Columbia Press, 2007.

———. "Xu Guangqi in His Context: The World of the Shanghai Gentry." In *Statecraft and Intellectual Renewal in Late Ming China: The Cross-Cultural Synthesis of Xu Guangqi (1562–1633),* ed. Catherine Jami, Peter Engelfriet, and Gregory Blue, 72–98. Leiden: Brill, 2001.

Brook, Timothy, and Gregory Blue, eds. *China and Historical Capitalism: Genealogies of Sinological Knowledge.* Cambridge: Cambridge University Press, 1999.

Brook, Timothy, Jérôme Bourgon, and Gregory Blue. *Death by a Thousand Cuts.* Cambridge, Mass.: Harvard University Press, 2008.

Buck, Pearl, trans. *All Men Are Brothers.* New York: J. Day, 1933.

Butlin, Ron. *The Exquisite Instrument.* Edinburgh: Salamander, 1982.

Cahill, James. *The Compelling Image: Nature and Style in Seventeenth-Century Chinese Painting*. Cambridge, Mass.: Harvard University Press, 1982.

———. *The Distant Mountains: Chinese Painting of the Late Ming Dynasty, 1570–1644*. Tokyo: Weatherhill, 1982.

———. *Parting at the Shore: Chinese Painting of the Early and Middle Ming Dynasty, 1368–1580*. Tokyo: Weatherhill, 1978.

Cao Shuji. *Zhongguo renkou shi* (Demographic history of China), vol. 4: *Ming shiqi* (The Ming period). Shanghai: Fudan daxue chubanshe, 2000.

Carswell, John. *Blue and White: Chinese Porcelain around the World*. Chicago: Art Media Resources, 2000.

Caviedes, César. *El Niño in History: Storming through the Ages*. Gainesville: University Press of Florida, 2001.

Chan, Albert. *The Glory and Fall of the Ming Dynasty*. Norman: University of Oklahoma Press, 1982.

Chan, Hok-lam. "Liu Ping-chung." In *In the Service of the Khan: Eminent Personalities of the Early Mongol-Yüan Period (1200–1300)*, ed. Igor de Rachewiltz, Hok-lan Chan, Hsiao Ch'i-ch'ing, and Peter Geier, 245–269. Wiesbaden: Harrassowitz Verlag, 1993.

———, and Wm. Theodore de Bary, eds. *Yüan Thought: Chinese Thought and Religion under the Mongols*. New York: Columbia University Press, 1982.

Chang, Kang-i Sun. *The Late Ming Poet Chen Tzu-lung: Crises of Love and Loyalism*. New Haven: Yale University Press, 1991.

Chang, Kang-i Sun, and Haun Saussy, eds. *Women Writers of Traditional China: An Anthology of Poetry and Criticism*. Stanford: Stanford University Press, 1999.

Chen Wenshi. "Mingdai weisuo de junhu" (Military households in the Ming guard system). Reprinted in *Mingshi yanjiu luncong*, ed. Wu Zhihe, vol. 2, 223–262. Taipei: Dali chubanshe, 1982.

Chia, Lucille. *Printing for Profit: The Commercial Publishers of Jianyang, Fujian (11th–17th Centuries)*. Cambridge, Mass.: Harvard University Asia Center, 2002.

Ching, Dora. "Visual Images of Zhu Yuanzhang." In *Long Live the Emperor!* ed. Sarah Schneewind, 171–209. Minneapolis: Society for Ming Studies, 2008.

Chow, Kai-wing. *Printing, Culture, and Power in Early Modern China*. Stanford: Stanford University Press, 2004.

Chu, Pingyi. "Trust, Instruments, and Cross-Cultural Scientific Exchanges: Chinese Debates over the Shape of the Earth, 1600–1800." *Science in Context* 12, 3 (1999): 385–411.

Church, Sally. "The Colossal Ships of Zheng He: Image or Reality?" In *Zheng He: Images and Perceptions*, ed. Roderich Ptak and Claudine Salmon, 156–176. Wiesbaden: Harrassowitz, 2005.

Clapp, Anne. *The Painting of T'ang Yin.* Chicago: University of Chicago Press, 1991.

Cleary, Thomas. *The Flower Ornament Scripture: A Translation of the Avatamsaka Sutra.* Boston: Shambhala, 1993.

Clunas, Craig. *Chinese Furniture.* London: Bamboo, 1988.

———. *Elegant Debts: The Social Art of Wen Zhengming.* London: Reaktion, 2004.

———. *Empire of Great Brightness: Visual and Material Cultures of Ming China 1368–1644.* London: Reaktion, 2007.

———. *Fruitful Sites: Garden Culture in Ming Dynasty China.* London: Reaktion, 1996.

———. *Superfluous Things: Material Culture and Social Status in Early Modern China.* Cambridge: Polity, 1991.

Coblin, W. South. "A Brief History of Mandarin." *Journal of the American Oriental Society* 120, 4 (Oct.–Dec. 2000): 537–552.

Cooper, Michael. *Rodrigues the Interpreter: An Early Jesuit in Japan and China.* Tokyo: Weatherhill, 1974.

Dai Mingshi. *Yulin chengshou jilüe* (Brief account of the defense of Yulin). Reprinted in *Dongnan jishi (wai shier zhong)* (A record of the southeast, plus twelve other records), ed. Shao Tingcai. Beijing: Zhonghua shuju, 2002.

Dardess, John. *A Ming Society: T'ai-ho County, Kiangsi, in the Fourteenth to Seventeenth Centuries.* Berkeley: University of California Press, 1996.

———. *Blood and History in China: The Donglin Faction and Its Repression, 1620–1627.* Honolulu: University of Hawaii Press, 2002.

———. "Did the Mongols Matter? Territory, Power, and the Intelligentsia in China from the Northern Song to the Early Ming." In *The Song-Yuan-Ming Transition in Chinese History,* ed. Paul Jakov Smith and Richard von Glahn, 111–134. Cambridge, Mass.: Harvard University Asia Center, 2003.

David, Percival, trans. *Chinese Connoisseurship: The Ko Ku Yao Lun, the Essential Criteria of Antiquities.* New York: Praeger, 1971.

de Bary, Theodore, trans. *Waiting for the Dawn: A Plan for the Prince: Huang Tsung-hsi's Ming-i-tai-fang lu.* New York: Columbia University Press, 1993.

de Rachewiltz, Igor; Hok-lan Chan; Hsiao Ch'i-ch'ing; and Peter Geier, eds. *In the Service of the Khan: Eminent Personalities of the Early Mongol-Yüan Period (1200–1300).* Wiesbaden: Harrassowitz Verlag, 1993.

Delgado, James. *Khubilai Khan's Lost Fleet: In Search of a Legendary Armada.* Vancouver: Douglas and McIntyre, 2008.

Deng, Kent. *The Premodern Chinese Economy: Structural Equilibrium and Capitalist Sterility.* London: Routledge, 1999.

Des Forges, Roger. *Cultural Centrality and Political Change in Chinese History:*

Northeast Henan in the Fall of the Ming. Stanford: Stanford University Press, 2003.

Di Cosmo, Nicola. "State Formation and Periodization in Inner Asian History." *Journal of World History* 10, 1 (1999): 1–40.

Dreyer, Edward. *Early Ming China: A Political History, 1355–1435*. Stanford: Stanford University Press, 1982.

Dudink, Ad. "Christianity in Late Ming China: Five Studies." Ph.D. diss., University of Leiden, 1995.

Dunstan, Helen. "The Late Ming Epidemics: A Preliminary Survey." *Ch'ing-shih wen-t'i* 23, 3 (November 1975): 1–59.

Ebrey, Patricia, trans. *Chu Hsi's Family Rituals: A Twelfth-Century Chinese Manual for the Performance of Cappings, Weddings, Funerals, and Ancestral Rites*. Princeton: Princeton University Press, 1991.

Elliott, Mark. "Hushuo: The Northern Other and *Han* Ethnogenesis." *China Heritage Quarterly* 19 (September 2009).

Elman, Benjamin. *A Cultural History of Civil Examinations in Late Imperial China*. Berkeley: University of California Press, 2000.

———. *On Their Own Terms: Science in China, 1550–1900*. Cambridge, Mass.: Harvard University Press, 2005.

Elvin, Mark. *The Retreat of the Elephants: An Environmental History of China*. New Haven: Yale University Press, 2004.

———. "Who Was Responsible for the Weather? Moral Meteorology in Late Imperial China." *Osiris*, 13 (1998): 213–237.

Farmer, Edward. *Early Ming Government: The Evolution of Dual Capitals*. Cambridge, Mass.: Harvard University Press, 1976.

———. *Zhu Yuanzhang and Early Ming Legislation: The Reordering of Chinese Society Following the Era of Mongol Rule*. Leiden: Brill, 1995.

Fei, Si-yen. *Negotiating Urban Space: Urbanization and Late Ming Nanjing*. Cambridge, Mass.: Harvard University Press, 2009.

Feng Yuejian. "Mingdai Jing-Hang yunhe de gongcheng guanli" (The management of the Grand Canal from the capital to Hangzhou in the Ming dynasty). *Zhongguo shi yanjiu*, 1993, 1: 50–60.

Fisher, Carney. *The Chosen One: Succession and Adoption in the Court of Ming Shizong*. Sydney: Allen and Unwin, 1990.

Fletcher, Joseph. "The Mongols: Ecological and Social Perspectives." *Harvard Journal of Asiatic Studies* 46 (1986): 11–50. Reprinted in his *Studies on Chinese and Islamic Inner Asia*, ed. Beatrice Forbes Manz. Farnham, Surrey: Ashgate, 1995.

Frank, Andre Gunder. *ReOrient: Global Economy in the Asian Age*. Berkeley: University of California Press, 1998.

Franke, Herbert. "Women under the Dynasties of Conquest." In *China under Mongol Rule*, ch. 6. London: Variorum, 1994.

———, and Denis Twitchett, eds. *The Cambridge History of China*, vol. 6: *Alien*

Regimes and Border States 907–1368. Cambridge: Cambridge University Press, 1994.

Furth, Charlotte. *A Flourishing Yin: Gender in China's Medical History, 960–1665*. Berkeley: University of California Press, 1998.

Gallagher, Louis, ed. *China in the Sixteenth Century: The Journals of Matthew Ricci, 1583–1610*. New York: Random House, 1953.

Gao Meiqing. *Paintings of the Ming Dynasty from the Palace Museum*. Hong Kong: The Chinese University of Hong Kong, 1988.

Geiss, James. "The Cheng-te Reign." In *The Cambridge History of China*, vol. 7: *The Ming Dynasty*, pt. 1, ed. Frederick Mote and Denis Twitchett, 403–439. Cambridge: Cambridge University Press, 1988.

———. "The Chia-ching Reign, 1522–1566." In *The Cambridge History of China*, vol. 7, ed. Frederick Mote and Denis Twitchett, 440–510. Cambridge: Cambridge University Press, 1988.

Girard, Pascale, ed. *Le Voyage en Chine d'Adriano de las Cortes S.J. (1625)*. Paris: Chandeigne, 2001.

Goodrich, L. Carrington, and Chaoying Fang, eds. *Dictionary of Ming Biography*. 2 vols. New York: Columbia University Press, 1976.

Gould, Stephen Jay. "Foreword" to Claudine Cohen, *The Fate of the Mammoth: Fossils, Myth, and History*. Chicago: University of Chicago Press, 1994.

Grove, Jean. "The Onset of the Little Ice Age." In *History and Climate: Memories of the Future?* ed. P. D. Jones, A. E. J. Ogilvie, T. D. Davies, and K. R. Briffa, 153–185. New York: Kluwer/Plenum, 2001.

Gu Gongxu, ed. *Catalogue of Chinese Earthquakes (1831 BC–1969 AD)*. Beijing: Science Press, 1989.

Guo Hong and Jin Runcheng. *Zhongguo xingzheng quhua tongshi: Mingdai juan* (A history of administrative jurisdictions in China: Ming volume). Shanghai: Fudan daxue chubanshe, 2007.

Handlin, Joanna. *Action in Late Ming Thought: The Reorientation of Lü Kun and Other Scholar-Officials*. Berkeley: University of California Press, 1983.

Hanson, Marta. "Inventing a Tradition in Chinese Medicine: From Universal Canon to Local Medical Knowledge in South China, the Seventeenth to the Nineteenth Century." Ph.D. diss., University of Pennsylvania, 1997.

Hartley, J. B., and David Woodward, eds. *Cartography in the Traditional East and Southeast Asian Societies*. Vol. 2, bk. 2 of *The History of Cartography*. Chicago: University of Chicago Press, 1994.

Hashimoto, Keizo. *Hsü Kuang-ch'i and Astronomical Reform: The Process of the Chinese Acceptance of Western Astronomy, 1629–1635*. Osaka: Kansai University Press, 1988.

Hazelton, Keith. "Patrilines and the Development of Localized Lineages: The Wu of Hsiu-ning City, Huichou, to 1528." In *Kinship Organization in Late Imperial China, 1000 to 1940*, ed. Patricia Ebrey and James Watson, 137–169. Berkeley: University of California Press, 1986.

Heijdra, Martin. "The Socio-Economic Development of Rural China during the Ming." In *The Cambridge History of China*, vol. 8: *The Ming Dynasty*, pt. 2, ed. Denis Twitchett and Frederick Mote, 417–578. Cambridge: Cambridge University Press, 1998.

Ho, Ping-ti. *Studies on the Population of China, 1368–1953*. Cambridge, Mass.: Harvard University Press, 1959.

Hoshi Ayao. *Chūgoku no shakai fukushi no rekishi* (The history of social welfare in China). Tokyo: Yamagawa shuppansha, 1988.

———. *Min-Shin jidai kōtsūshi no kenkyū* (Studies in the transportation history of the Ming-Qing period). Tokyo: Yamakawa, 1971.

Hsia, Ronnie. *A Jesuit in the Forbidden City: Matteo Ricci, 1552–1610*. Oxford: Oxford University Press, 2010.

Hsiao, Ch'i-ch'ing. "Mid-Yüan Politics." In *The Cambridge History of China*, vol. 6, ed. Herbert Franke and Denis Twitchett, 490–560. Cambridge: Cambridge University Press, 1994.

Hsü, Ginger. *A Bushel of Pearls: Paintings for Sale in Eighteenth-Century Yangzhou*. Stanford: Stanford University Press, 2001.

Huang Qinglian. *Yuandai huji zhidu yanjiu* (Studies in the household registration system of the Yuan dynasty). Taipei: Guoli Taiwan daxue wenxuebu, 1977.

Huang, Ray. *1587, a Year of No Significance*. New Haven: Yale University Press, 1981.

———. "The Lung-ch'ing and Wan-li Reigns, 1567–1620." In *The Cambridge History of China*, vol. 7: *The Ming Dynasty*, pt. 1, ed. Frederick Mote and Denis Twitchett, 511–584. Cambridge: Cambridge University Press, 1988.

———. "The Ming Fiscal Administration." In *The Cambridge History of China*, vol. 8: *The Ming Dynasty*, pt. 2, ed. Denis Twitchett and Frederick Mote, 106–171. Cambridge: Cambridge University Press, 1998.

———. *Taxation and Governmental Finance in Sixteenth-Century Ming China*. Cambridge: Cambridge University Press, 1974.

Huang Yi-Long. "Sun Yuanhua (1581–1632): A Christian Convert Who Put Xu Guangqi's Military Reform Policy into Practice." In *Statecraft and Intellectual Renewal in Late Ming China: The Cross-Cultural Synthesis of Xu Guangqi (1562–1633)*, ed. Catherine Jami, Peter Engelfriet, and Gregory Blue, 225–259. Leiden: Brill, 2001.

Hucker, Charles. *A Dictionary of Official Titles in Imperial China*. Stanford: Stanford University Press, 1985.

———. *The Ming Dynasty: Its Origins and Evolving Institutions*. Ann Arbor: Center for Chinese Studies, University of Michigan, 1978.

———. "Ming Government." In *The Cambridge History of China*, vol. 8: *The Ming Dynasty*, pt. 2, ed. Denis Twitchett and Frederick Mote, 9–105. Cambridge: Cambridge University Press, 1998.

Hurn, Samantha. "Here Be Dragons? No, Big Cats." *Anthropology Today* 25, 1 (February 2009): 6–11.

Jami, Catherine, Peter Engelfriet, and Gregory Blue, eds. *Statecraft and Intellectual Renewal in Late Ming China: The Cross-Cultural Synthesis of Xu Guangqi (1562–1633).* Leiden: Brill, 2001.

Jay, Jennifer. *A Change in Dynasties: Loyalism in Thirteenth-Century China.* Bellingham: Center for East Asian Studies, Western Washington University, 1991.

Jiang, Yonglin, trans. *The Great Ming Code.* Seattle: University of Washington Press, 2005.

Johnston, Iain. *Cultural Realism: Strategic Culture and Grand Strategy in Chinese History.* Princeton: Princeton University Press, 1995.

Kieschnick, John. *The Impact of Buddhism on Chinese Material Culture.* Princeton: Princeton University Press, 2003.

Knapp, Robert. *Chinese Landscapes: The Village as Place.* Honolulu: University of Hawaii Press, 1992.

Kuroda Akinobu. "Copper Coins Chosen and Silver Differentiated." *Acta Asiatica* 88 (2005), 65–86.

Kutcher, Norman. *Mourning in Late Imperial China: Filial Piety and the State.* Cambridge: Cambridge University Press, 1999.

Lau, D. C., trans. *Mencius.* London: Penguin, 1970.

Lamb, H. H. *Climate: Present, Past and Future,* vol. 2: *Climatic History and the Future.* Princeton: Princeton University Press, 1985.

Ledyard, Gari. "Cartography in Korea." In *Cartography in the Traditional East and Southeast Asian Societies,* ed. J. B. Hartley and David Woodward, 235–345. Chicago: University of Chicago Press, 1994.

Legge, James, trans. *The Confucian Classics,* vol. 1. Oxford: Clarendon Press, 1893.

Lentz, Harris. *The Volcano Registry: Names, Locations, Descriptions and History for over 1500 Sites.* Jefferson, NC: McFarland, 1999.

Lewis, Mark Edward. "The Mythology of Early China." In *Early Chinese Religion,* vol. 1, pt. 1, ed. John Lagerway and Marc Kalinowski, 543–594. Leiden: Brill, 2009.

Li Bozhong. "Was There a 'Fourteenth-Century Turning Point'? Population, Land, Technology, and Farm Management." In *The Song-Yuan-Ming Transition in Chinese History,* ed. Paul Jakov Smith and Richard von Glahn, 135–175. Cambridge, Mass.: Harvard University Asia Center, 2003.

Li Chu-tsing. "Li Rihua and His Literati Circle in the Late Ming Dynasty." *Orientations* 18, 8 (August 1987): 28–39.

——— et al. *The Chinese Scholar's Studio: Artistic Life in the Late Ming Period.* New York: Thames and Hudson, 1987.

Li Defu. *Mingdai renkou yu jingji fazhan* (Population and economic development in the Ming dynasty). Beijing: Zhongguo shehui kexue chubanshe, 2008.

Li, He, and Michael Knight. *Power and Glory: Court Arts of China's Ming Dynasty.* San Francisco: Asian Art Museum, 2008.

Li Xiaocong. *Ouzhou shoucang bufen Zhongwen gu ditu xulu* (A descriptive catalogue of pre-1900 Chinese maps as seen in Europe). Beijing: Guoji wenhua chuban gongsi, 1996.

Liang Fangzhong, ed. *Zhongguo lidai hukou, tiandi, tianfu tongji* (Chinese population, land, and taxation statistics for successive dynasties). Shanghai: Shanghai renmin chubanshe, 1980.

Lieberman, Victor. *Strange Parallels: Southeast Asia in Global Context, c. 800–1830.* New York: Cambridge University Press, 2002.

Littrup, Leif. *Subbureaucratic Government in China in Ming Times: A Study of Shandong Province in the Sixteenth Century.* Oslo: Universitetsforlaget, 1981.

Liu Cuirong. *Ming-Qing shiqi jiazu renkou yu shehui jingji bianqian* (Family demography and socioeconomic change in the Ming-Qing period). Taipei: Zhongyang yanjiuyuan jingji yanjiusuo, 1992.

Liu Ts'un-yan and Judith Berling. "The 'Three Teachings' in the Mongol-Yüan Period." In *Yüan Thought: Chinese Thought and Religion under the Mongols,* ed. Hok-lan Chan and Wm. Theodore de Bary, 479–512. New York: Columbia University Press, 1982.

Ma Zhibing. "Mingchao tudi fazhi" (Land law in the Ming dynasty). In *Zhonggou lidai tudi ziyuan fazhi yanjiu* (Studies in the law on land resources in China through the dynasties), ed. Pu Jian, 405–458. Beijing: Beijing daxue chubanshe, 2006.

Marks, Robert. "China's Population Size during the Ming and Qing: A Comment on the Mote Revision." Paper presented at the annual meeting of the Association for Asian Studies, 2002.

———. *Tigers, Rice, Silk, and Silt: Environment and Economy in Late Imperial South China.* Cambridge: Cambridge University Press, 1997.

McDermott, Joseph. *A Social History of the Chinese Book: Books and Literati Culture in Late Imperial China.* Hong Kong: Hong Kong University Press, 2006.

McNeill, William. *Plagues and Peoples.* Harmondsworth: Penguin, 1979.

Menegon, Eugenio. *Ancestors, Virgins, and Friars: Christianity as a Local Religion in Late Imperial China.* Cambridge, Mass.: Harvard University Press, 2009.

Menzies, Nicholas. "Forestry." In *Science and Civilisation in China* VI:3, ed. Joseph Needham, 539–667. Cambridge: Cambridge University Press, 1996.

Meskill, John, trans. *Ch'oe Pu's Diary: A Record of Drifting across the Sea.* Tucson: University of Arizona Press, 1965.

Mote, Frederick. "Chinese Society under Mongol Rule, 1215–1368." In *The*

Cambridge History of China, vol. 6, ed. Herbert Franke and Denis Twitchett, 616–664.

———. "The Growth of Chinese Despotism: A Critique of Wittfogel's Theory of Oriental Despotism as Applied to China." *Oriens Entremus* 8 (1961): 1–41.

Mote, Frederick, and Denis Twitchett, eds. *The Cambridge History of China,* vol. 7: *The Ming Dynasty 1368–1644,* pt. 1. Cambridge: Cambridge University Press, 1988.

Moule, A. C., and Paul Pelliot, trans. *The Description of the World.* London: Routledge, 1938.

Nappi, Carla. *The Monkey and the Inkpot: Natural History and Its Transformation in Early Modern China.* Cambridge, Mass.: Harvard University Press, 2009.

Needham, Joseph, and Robin Yates. *Science and Civilisation in China* V:6 (Military Technology: Missiles and Sieges). Cambridge: Cambridge University Press, 1994.

Nimick, Thomas. *Local Administration in Ming China: The Changing Roles of Magistrates, Prefects, and Provincial Officials.* Minneapolis: Society for Ming Studies, 2008.

Oertling, Sewall. *Painting and Calligraphy in the Wu-tsa-tsu.* Ann Arbor: Center for Chinese Studies, University of Michigan, 1997.

Parsons, James. "The Ming Dynasty Bureaucracy: Aspects of Background Forces." In *Chinese Government in Ming Times: Seven Studies,* ed. Charles Hucker, 175–232. New York: Columbia University Press, 1969.

———. *The Peasant Rebellions of the Late Ming Dynasty.* Tucson: University of Arizona Press, 1970.

Peng Xinwei. *Zhongguo huobi shi* (A history of Chinese currency). Beijing: Qunlian chubanshe, 1954.

Peterson, Willard. "Why Did They Become Christians? Yang Tingyun, Li Zhizao, and Xu Guangqi." In *East Meets West: The Jesuits in China, 1582–1773,* ed. Charles Ronan and Bonnie Oh. Chicago: Loyola University Press, 1988.

Plaks, Andrew. *The Four Masterworks of the Ming Novel.* Princeton: Princeton University Press, 1987.

Polo, Marco. *The Travels,* trans. Ronald Latham. Harmondsworth: Penguin, 1958.

Pomeranz, Kenneth. *The Great Divergence: China, Europe, and the Making of the Modern World Economy.* Princeton: Princeton University Press, 2000.

Ptak, Roderich, and Claudine Salmon, eds. *Zheng He: Images and Perceptions.* South China and Maritime Asia, vol. 15. Wiesbaden: Harrassowitz, 2005.

Qiu Zhonglin. "Mingdai changcheng yanxian de zhimu zaolin" (Afforestation along the Great Wall in the Ming dynasty). *Nankai daxue xuebao* 2007, 3: 32–42.

Quinn, William. "A Study of Southern Oscillation-Related Climatic Activity for A.D. 633–1900 Incorporating Nile River Flood Data." In *El Niño: Historical and Paleoclimatic Aspects of the Southern Oscillation*, ed. Henry Diaz and Vera Markgraf, 119–149. Cambridge: Cambridge University Press, 1992.

Ratchnevsky, Paul. *Genghis Khan: His Life and Legacy.* Oxford: Blackwell, 1991.

Reid, Anthony. *Southeast Asia in the Age of Commerce, 1458–1680,* vol. 2: *Expansion and Crisis.* New Haven: Yale University Press, 1993.

Riello, Giorgio, and Prasannan Parthasarathi, eds. *The Spinning World: A Global History of Cotton Textiles.* Oxford: Oxford University Press, 2009.

Rosenblatt, Jason. *Renaissance England's Chief Rabbi.* Oxford: Oxford University Press, 2006.

Rossabi, Morris. *Khubilai Khan: His Life and Times.* Berkeley: University of California Press, 1988.

———. *Voyager from Xanadu: Rabban Sauma and the First Journey from China to the West.* Tokyo: Kodansha, 1992.

Rowe, William. *Hankow: Commerce and Society in a Chinese City, 1796–1889.* Stanford: Stanford University Press, 1984.

Roy, David, trans. *Plum in the Golden Vase.* 3 vols. Princeton: Princeton University Press, 1993.

Ruitenbeek, Klaas. *Carpentry and Building in Late Imperial China: A Study of the Fifteenth-Century Carpenter's Manual Lu Ban Jing.* Leiden: E. J. Brill, 1993.

Salmon, Claudine, and Roderich Ptak, eds. *Zheng He: Images and Perceptions/Bilder und Wahrnehmingen.* Wiesbaden: Harrassowitz, 2005.

Schäfer, Dagmar, and Dieter Kuhn. *Weaving an Economic Pattern in Ming Times (1368–1644): The Production of Silk Weaves in the State-Owned Silk Workshops.* Würzburger Sinologische Schriften. Heidelberg: Forum, 2002.

Schmalzer, Sigrid. *The People's Peking Man: Popular Science and Human Identity in Twentieth-Century China.* Chicago: University of Chicago Press, 2008.

Schneewind, Sarah. *Community Schools and the State in Ming China.* Stanford: Stanford University Press, 2006.

———, ed. *Long Live the Emperor! Uses of the Ming Founder across Six Centuries of East Asian History.* Minneapolis: Society for Ming Studies, 2008.

Scott, James C. *The Art of Not Being Governed: An Anarchist History of Upland Southeast Asia.* New Haven: Yale University Press, 2009.

Sedo, Timothy. "Environmental Jurisdiction within the Mid-Ming Yellow River

Flood Plain." Paper presented at the annual meeting of the Association for Asian Studies, 2008.

Shin, Leo. *The Making of the Chinese State: Ethnicity and Expansion on the Ming Borderlands.* Cambridge: Cambridge University Press, 2006.

Shinno, Reiko. "Medical Schools and the Temples of the Three Progenitors in Yuan China: A Case of Cross-Cultural Interactions." *Harvard Journal of Asiatic Studies* 67, 1 (June 2007): 89–133.

Smith, Paul Jakov, and Richard von Glahn, eds. *The Song-Yuan-Ming Transition in Chinese History.* Cambridge, Mass.: Harvard University Asia Center, 2003.

So, K. L. Billy. *Prosperity, Region, and Institutions in Maritime China: The South Fukien Pattern, 946–1368.* Cambridge, Mass.: Harvard University Asia Center, 2000.

Spence, Jonathan. *The Memory Palace of Matteo Ricci.* Harmondsworth: Penguin, 1985.

———. *Return to Dragon Mountain: Memories of a Late Ming Man.* New York: Viking, 2007.

Standaert, Nicolas, ed. *Handbook of Christianity in China,* vol. 1: 635–1800. Leiden: Brill, 2001.

———. *Yang Tingyun, Confucian and Christian in Late Ming China: His Life and Thought.* Leiden: E. J. Brill, 1988.

Sterckx, Roel. *The Animal and the Daemon in Early China.* Albany: State University of New York Press, 2002.

Struve, Lynn. *The Ming-Qing Conflict, 1619–1683: A Historiography and Source Guide.* Ann Arbor: Association for Asian Studies, 1998.

———. *The Southern Ming 1644–1662.* New Haven: Yale University Press, 1984.

———, ed. *Time, Temporality, and Imperial Transition: East Asia from Ming to Qing.* Honolulu: University of Hawaii Press, 2005.

———. *Voices from the Ming-Qing Cataclysm: China in Tigers' Jaws.* New Haven: Yale University Press, 1993.

Stuart-Fox, Martin. "Mongol Expansionism." Reprinted in *China and Southeast Asia,* ed. Geoff Wade, vol. 1, 365–378. London: Routledge, 2009.

Subrahmanyam, Sanjay. *The Portuguese Empire in Asia, 1500–1700: A Political and Economic History.* New York: Longman, 1993.

Sun, Jinghao. "City, State, and the Grand Canal: Jining's Identity and Transformation, 1289–1937." Ph.D. diss., University of Toronto, 2007.

Sun, Laichen. "Ming-Southeast Asian Overland Interactions, 1368–1644." Ph.D. diss., University of Michigan, 2000.

Sung Ying-hsing. *T'ien-kung k'ai-wu; Chinese Technology in the Seventeenth Century,* trans. E-tu Zen Sun and Shiou-chuan Sun. University Park: Pennsylvania State University Press, 1966.

Szonyi, Michael. "The Cult of Hu Tianbao and the Eighteenth-Century Discourse of Homosexuality." *Late Imperial China* 19, 1 (June 1998): 1–25.

Taga Akigorō. *Chūgoku sōfu no kenkyū* (Studies in Chinese lineage genealogies). Tokyo: Nihon gakujutsu shinkōkai, 1981.

Tian Rukang (T'ien Ju-K'ang). "*Duhai fangcheng*—Zhongguo diyiben keyin de shuilupu" (The first printed Chinese rutter—*Duhai fangcheng*). In *Zhongguo kejishi tansuo* (Explorations in the history of science and technology in China), ed. Li Guohao, Zhang Mengwen, and Cao Tianqin, 301–308. Shanghai: Shanghai Chinese Classics Publishing House, 1982.

Tong, James. *Disorder under Heaven: Collective Violence in the Ming Dynasty.* Stanford: Stanford University Press, 1991.

Topsell, Edward. *The Historie of Serpents.* London: William Jagger, 1608.

Trevor-Roper, H. R. *Archbishop Laud, 1573–1645*, 2nd ed. London: Macmillan, 1963.

Tsai, Shih-shan Henry. *The Eunuchs in the Ming Dynasty.* Albany: State University of New York Press, 1996.

———. *Perpetual Happiness: The Ming Emperor Yongle.* Seattle: University of Washington Press, 2001.

Tu, Wei-ming. *Neo-Confucian Thought in Action: Wang Yang-ming's Youth (1572–1509).* Berkeley: University of California Press, 1976.

Twitchett, Denis, and Tileman Grimm. "The Cheng-t'ung, Ching-t'ai, and T'ien-shun Reigns, 1436–1464." In *The Cambridge History of China,* vol. 7: *The Ming Dynasty,* pt. 1, ed. Frederick Mote and Denis Twitchett, 305–342. Cambridge: Cambridge University Press, 1988.

Twitchett, Denis and Frederick Mote, eds. *The Cambridge History of China,* vol. 8: *The Ming Dynasty 1368–1644,* pt. 2. Cambridge: Cambridge University Press, 1997.

Unschuld, Paul. *Medicine in China: A History of Pharmaceutics.* Berkeley: University of California Press, 1986.

Von Glahn, Richard. *Fountain of Fortune; Money and Monetary Policy in China, 1000–1700,* Berkeley: University of California Press, 1996.

Wade, Geoff. *Southeast Asia in the Ming Shi-lu: An Open Access Resource.* Singapore: Asia Research Institute and the Singapore E-Press, National University of Singapore, accessed 2010.

———. "The Zheng He Voyages: A Reassessment." Reprinted in *China and Southeast Asia,* vol. 2, 118–141. London: Routledge, 2009.

Wakeman, Frederic, Jr. *The Great Enterprise: The Manchu Reconstruction of Imperial Order in Seventeenth-Century China.* 2 vols. Berkeley: University of California Press, 1985.

Waldron, Arthur. *The Great Wall: From History to Myth.* New York: Cambridge University Press, 1990.

Waley, Arthur, trans. *The Analects of Confucius*. London: Allen and Unwin, 1949.

———, trans. *Monkey*. New York: Grove, 1970.

Wallerstein, Immanuel. *The Modern World-System II: Mercantilism and the Consolidation of the European World-Economy, 1600–1750*. New York: Academic Press, 1980.

Waltner, Ann. *Getting an Heir: Adoption and the Construction of Kinship in Late Imperial China*. Honolulu: University of Hawaii Press, 1990.

Wang Shao-wu and Wei Gao. "La Niña and Its Impacts on China's Climate." In *La Niña and Its Impacts: Facts and Speculation*, ed. Michael Glantz, 186–189. Tokyo: United Nations University Press, 2002.

Wang Shuanghuai and Fang Jun, eds. *Zhonghua rili tongdian* (A complete Chinese calendar), vol. 4: *Yuan Ming Qing rili* (Calendar of the Yuan, Ming, and Qing). Changchun: Jilin wenshi chubanshe, 2006.

Wang Yong. *Zhongguo ditu shigang* (An outline history of Chinese maps). Beijing, 1958.

Wang Yü-ch'üan. "The Rise of Land Tax and the Fall of Dynasties in Chinese History." *Pacific Affairs* 9, 2 (June 1936).

Watt, James C. Y., and Denise Patry Leidy. *Defining Yongle: Imperial Art in Early Fifteenth-Century China*. New York: Metropolitan Museum of Art, 2005.

Weitz, Ankeney. *Zhou Mi's Record of Clouds and Mist Passing before One's Eyes: An Annotated Translation*. Leiden: Brill, 2002.

Will, Pierre-Étienne, and R. Bin Wong, eds. *Nourish the People: The State Civilian Granary System in China, 1650-1850*. Ann Arbor: Center for Chinese Studies, University of Michigan, 1991.

Wong, R. Bin. *China Transformed: Historical Change and the Limits of European Experience*. Ithaca: Cornell University Press, 1997.

Wood, Frances. *Did Marco Polo Go to China?* London: Secker and Warburg, 1995.

Woodside, Alexander. "The Ch'ien-lung Reign." In *The Cambridge History of China*, vol. 9, ed. Willard Peterson, 230–309. Cambridge: Cambridge University Press, 2002.

Wu Chengluo. *Zhongguo duliangheng shi* (A history of Chinese measurements). Shanghai: Shangwu yinshuguan, 1957.

Wu Han. *Jiang Zhe cangshujia shilüe* (A brief history of book collectors in Jiangsu and Zhejiang). Reprint, Beijing: Zhonghua shuju, 1981.

Wu Hongqi. *Yuandai nongye dili* (Economic geography of the Yuan dynasty). Xi'an: Xi'an ditu chubanshe, 1997.

Wu Jihua. *Mingdai haiyun ji yunhe de yanjiu* (Sea transport and the Grand Canal in the Ming dynasty). Taipei: Academia Sinica, 1961.

Wu, Marshall. *The Orchid Pavilion Gathering: Chinese Painting from the University of Michigan Museum of Art.* 2 vols. Ann Arbor: University of Michigan, 2000.

Wu Renshu. *Pinwei shehua: Wan Ming de xiaofei shehui yu shidafu* (Delicacy and extravagance: consumer society and the gentry in the late Ming). Taipei: Lianjing chubanshe, 2007.

Xiamen daxue lishixi (History department of Amoy University), ed. *Li Zhi yanjiu cankao ziliao* (Reference materials for the study of Li Zhi). Xiamen.

Yang Zhengtai. "Ming-Qing Linqing de shengshuai yu dili tiaojian de bianhua" (The rise and fall of Linqing in the Ming and Qing in relation to changes in geographical conditions). *Lishi dili* 3 (1983): 115–120.

Yee, Cordell. "Reinterpreting Traditional Chinese Geographical Maps." In *Cartography in the Traditional East and Southeast Asian Societies,* ed. J. B. Hartley and David Woodward, 35–70. Chicago: University of Chicago Press, 1994.

———. "Taking the World's Measure: Chinese Maps between Observation and Text." In *Cartography in the Traditional East and Southeast Asian Societies,* ed. J. B. Hartley and David Woodward, 117–124. Chicago: University of Chicago Press, 1994.

Yü, Chün-fang. *The Renewal of Buddhism in China: Chu-hung and the Late Ming Synthesis.* New York: Columbia University Press, 1981.

Zhang Jiacheng and Thomas Crowley. "Historical Climate Records in China and Reconstruction of Past Climates." *Journal of Climate* 2 (August 1989): 833–849.

Zhang Qing. *Hongdong dahuaishu yimin zhi* (Migration gazetteer of the old locust tree in Hongdong). Taiyuan: Shanxi guji chubanshe, 2000.

Zhao Shiyu and Du Zhengzhen. "'Birthday of the Sun': Historical Memory in Southeastern Coastal China of the Chongzhen Emperor's Death." In *Time, Temporality, and Imperial Transition: East Asia from Ming to Qing,* ed. Lynn A. Struve, 244–276. Honolulu: University of Hawaii Press, 2005.

Zhongyang qixiangju qixiang kexue yanjiuyuan (Meteorological research institute of the central meteorological bureau), ed. *Zhongguo jin wubai nian hanlao fenbu tuji* (Atlas of the distribution of drought and wetness in China for the last five hundred years). Beijing: Ditu chubanshe, 1981.

Zhou Zhiyuan. *Mingdai huangzheng wenxian yanjiu* (Studies on famine administration texts of the Ming dynasty). Hefei: Anhui daxue chubanshe, 2007.

Zurndorfer, Harriet. "The Resistant Fibre: Cotton Textiles in Imperial China." In *The Spinning World: A Global History of Cotton Textiles, 1200–1850,* ed. Giorgio Riello and Prasannan Parthasarathi, 43–62. Oxford: Oxford University Press, 2009.

ACKNOWLEDGMENTS

My family put up with a great deal from me, but they were obliged to put up with more than the usual deal during the five months when I was writing this book. I failed to ask any of them for their indulgence at the time, and now ask their forgiveness for the neglect that writing imposed. Fay, Vanessa, Katie, Taylor, Jonah: I hope I am back in the range of normal.

I wrote the book during my last five months as principal of St. John's College at the University of British Columbia, and am grateful to the staff and fellows of the College for sustaining me in this project.

For introducing me to the Selden map and providing me with a reproduction of it, I am forever grateful to David Helliwell—and wish we were still colleagues. For reading portions of the manuscript as it was coming into being, often at short notice, I wish to thank Desmond Cheung, Fei Siyen, Noa Grass, Marta Hanson, Carla Nappi, Tim Sedo, and Chelsea Wang. For providing detailed page-by-page critiques, I am particularly grateful to Peter Ditmanson and Jim Wilkerson, and most of all to Fay Sims.

To Susan Wallace Boehmer at Harvard University Press, I offer my heartfelt thanks for the outstanding work she has done, and the extraordinary patience she has shown, in editing not just this volume but the entire set of six in the History of Imperial China. I wish to reserve my final acknowledgment, however, for Kathleen McDermott, the Press's Senior Editor for History, who originally proposed that I edit the series and gave me both the advice and the freedom I needed to write this volume in the way I have.

INDEX